Karl Dvorsky

Analysis of Nonlinear Heat Transfer in Electric Cables

Karl Dvorsky

Analysis of Nonlinear Heat Transfer in Electric Cables

Reduction of PDE Models and Solution Methods

Südwestdeutscher Verlag für Hochschulschriften

Impressum / Imprint
Bibliografische Information der Deutschen Nationalbibliothek: Die Deutsche Nationalbibliothek verzeichnet diese Publikation in der Deutschen Nationalbibliografie; detaillierte bibliografische Daten sind im Internet über http://dnb.d-nb.de abrufbar.
Alle in diesem Buch genannten Marken und Produktnamen unterliegen warenzeichen-, marken- oder patentrechtlichem Schutz bzw. sind Warenzeichen oder eingetragene Warenzeichen der jeweiligen Inhaber. Die Wiedergabe von Marken, Produktnamen, Gebrauchsnamen, Handelsnamen, Warenbezeichnungen u.s.w. in diesem Werk berechtigt auch ohne besondere Kennzeichnung nicht zu der Annahme, dass solche Namen im Sinne der Warenzeichen- und Markenschutzgesetzgebung als frei zu betrachten wären und daher von jedermann benutzt werden dürften.

Bibliographic information published by the Deutsche Nationalbibliothek: The Deutsche Nationalbibliothek lists this publication in the Deutsche Nationalbibliografie; detailed bibliographic data are available in the Internet at http://dnb.d-nb.de.
Any brand names and product names mentioned in this book are subject to trademark, brand or patent protection and are trademarks or registered trademarks of their respective holders. The use of brand names, product names, common names, trade names, product descriptions etc. even without a particular marking in this works is in no way to be construed to mean that such names may be regarded as unrestricted in respect of trademark and brand protection legislation and could thus be used by anyone.

Coverbild / Cover image: www.ingimage.com

Verlag / Publisher:
Südwestdeutscher Verlag für Hochschulschriften
ist ein Imprint der / is a trademark of
AV Akademikerverlag GmbH & Co. KG
Heinrich-Böcking-Str. 6-8, 66121 Saarbrücken, Deutschland / Germany
Email: info@svh-verlag.de

Herstellung: siehe letzte Seite /
Printed at: see last page
ISBN: 978-3-8381-3722-3

Zugl. / Approved by: München, Universität der Bundeswehr, Dissertation, 2012

Copyright © 2013 AV Akademikerverlag GmbH & Co. KG
Alle Rechte vorbehalten. / All rights reserved. Saarbrücken 2013

Acknowledgement

I would like to express my gratitude to Professor Joachim Gwinner for his advice. Especially his profound knowledge in monotone operator theory and boundary integral equations was very helpful. I would like to thank Professor Hans-Dieter Liess. It was him who motivated the subject of the book. I am grateful to my Colleagues at the University of the Bundeswehr München: Nina Ovcharova, Florian Loos, Daniel Mohr, Sven-Joachim Kimmerle and Martin Schlüter for the fruitful cooperation and inspiring discussions. I would also like to thank for the support of our secretary Mrs. Lößl.

Contents

1. Introduction   7

2. Asymptotic analysis of the initial boundary value problem   11
   2.1. Asymptotics for large times 11
      2.1.1. Setup of the initial boundary value problem 11
      2.1.2. Existence and uniqueness of a stationary solution . . . 13
      2.1.3. Treatment of the dynamical problem 18
      2.1.4. Analysis of (2.1) for constant temperature profiles . . . 21
   2.2. Approximation of subresonant solutions 28
      2.2.1. Sensitivity Results 29
      2.2.2. Error minimizing choice of the Poisson datum 30
   2.3. Asymptotic behaviour in cylindrical domains 31
      2.3.1. Setting of the boundary value problems under investigation 31
      2.3.2. Approximation of u_l by u_∞ 34
      2.3.3. Locality of the convergence 37
   2.4. An estimate for $c_\star$ in star-shaped domains 40
      2.4.1. Preliminary remarks 40
      2.4.2. Inhomogeneous Friedrichs inequality in $W^{1,p}(\Omega)$ 42
      2.4.3. Estimate for the trace embedding $W^{1,p}(\Omega) \hookrightarrow L^p(\Gamma_g)$. 47
      2.4.4. An extension to contractible finite path domains 49
   2.5. Combining of the estimates 51
      2.5.1. Setting of the general problem 51
      2.5.2. Setting of the reduced problem 52

| | | 2.5.3. | Combining estimate . . . . | 53 |

3. Estimates for heat transfer in electric cables   55

- 3.1. Estimates for a uninsulated cable 55
  - 3.1.1. Modelling of the heat transfer problem 55
  - 3.1.2. Identification of the general setting with physical quantities 58
  - 3.1.3. Subresonant states and long time behaviour 59
  - 3.1.4. Sensitivity for $\alpha_\rho \to 0$ and asymptotics for $l \to \infty$ . . . 64
  - 3.1.5. Application to physical data 66
  - 3.1.6. Oscillating behaviour of stationary solutions for large temperature coefficients α_ρ 70
- 3.2. Estimates for an insulated cable 72
  - 3.2.1. Modelling of the problem 72
  - 3.2.2. Subresonant states and long-time behaviour 74
  - 3.2.3. Transformation of the monotone boundary condition . 76
  - 3.2.4. Subresonance for the transformed problem (3.20) and its sensitivity and asymptotics for $\alpha_\rho \to 0$, $l \to \infty$ . . . 80
  - 3.2.5. Remarks on the transformation of problem (3.15) . . . 85

4. Treatment by nonlinear boundary integral equations   88

- 4.1. Boundary integral approach for uninsulated cables 89
  - 4.1.1. Setup of the problem 89
  - 4.1.2. Equivalent formulation by a nonlinear boundary integral equation 92
  - 4.1.3. Existence and Uniqueness of a solution of the nonlinear boundary integral equation 95
  - 4.1.4. Iterative determination of the boundary temperature as a fixed point 98
  - 4.1.5. Case of rotational symmetry 102
  - 4.1.6. Illustration of Theorem 4.2 107

4.2.	Boundary integral approach for insulated cables		109
	4.2.1.	Setup of the problem	109
	4.2.2.	The outer domain formulation	110
	4.2.3.	Determination of the heat flow	111
	4.2.4.	Boundary integral approach on doubly connected domains	114
	4.2.5.	Iterative determination of the boundary temperatures	119
	4.2.6.	The case of a multiply connected domain	123
	4.2.7.	The case of rotational symmetry	126
	4.2.8.	An application to physical data	130

5. Conclusions   **132**

A. Appendix   **134**

1. Introduction

The main motivation for this work is a lack of theoretical background in the field of heat transfer in electric cables. In engineering, there are several methods which reduce a full transient three dimensional problem to a more simple one. For large times, one common method is to reduce the time dependent problem to a stationary problem. Another heuristic method simplifies the three dimensional problem in a cylindrical domain to a cross-sectional two dimensional problem when the axial dimension of the cylinder becomes large. These, in engineering very common and useful reduction methods are often applied without knowledge of the resulting error. This again leads to rather nebulous criteria which shall decide if it is reasonable to use a specific simplification or not. Here we develop a heat transfer study where the particular reductions are treated via an asymptotic analysis of the associated parameters, time and length among others. Then we apply the asymptotic estimates to the specific setting of an electric cable and identify the associated abstract parameters with explicit physical quantities. Within the reduced model, we use nonlinear boundary integral methods applied to multiply connected domains. An iterative procedure computes the relevant temperatures on the boundary of the cross-sectional domain.

In chapter 2 we consider a semilinear parabolic boundary value problem with nonlinear boundary conditions. More precisely, we look for a function u depending on time and space that satisfies

$$u_t = \operatorname{div}(\Lambda \nabla u) + \varsigma\, r(u) + f \quad \text{in } \Omega \times (0, \infty)\, ;\ \varsigma \in \mathbb{R} \quad (1.1)$$
$$-(\Lambda \nabla u)\, n = \beta(u) \quad \text{on } \Gamma \times (0, \infty)\, ;\ u = u_{init} \quad \text{on } \Omega \times \{0\}\, .$$

Under appropriate assumptions on the data we will give an existence and uniqueness result for (1.1). It relies on a subresonance condition which ensures that the heat generating nonlinear function r in Ω is not too large compared to the heat emitting function β on the boundary Γ. Then, for $\Omega = \Omega_{cr} \times (-l, l)$, $\Gamma = \partial\Omega_{cr} \times (-l, l)$, $l > 0$, we reduce (1.1) stepwise with $t \to \infty$, $\varsigma \to 0$, $l \to \infty$ to the stationary problem of finding $\bar{u}$ on the cross section Ω_{cr}

$$-\operatorname{div}\left(\bar{\Lambda}\,\nabla\bar{u}\right) \;=\; \bar{f} \quad \text{in } \Omega_{cr} \qquad (1.2)$$
$$-\left(\bar{\Lambda}\,\nabla\bar{u}\right)n \;=\; \beta(\bar{u}) \quad \text{on } \partial\Omega_{cr}\,.$$

Hence we show a controlled, i.e. estimated, reduction of the full problem (1.1) to (1.2). One standard procedure for direct numerical solutions of the full problem are finite element and finite volume methods for parabolic problems with nonlinear boundary conditions, see e.g. [19] [54], [29]. The main advantage of these procedures lies in the accurate solution of the problem (1.1) at least in the well posed, i.e. subresonant, case.

Now there are several aspects which underline the advantage of the reduction of the full model over finite element-/finite volume-method in industrial applications. Firstly the essential input data - such as electrical current and conductor cross section area - are known just up to a certain tolerance which often exceeds 5%. Hence, in this context, the accuracy of the FE/FV-procedures maps the inaccuracy of the input data only. Secondly, the numerical procedures solving the reduced problems are faster by orders of magnitude compared to the numerical solution of the full problem. This enables an extensive variation of the input parameters to treat inverse probems. Above all, which geometry is appropriate if a certain current load should not exceed a critical temperature of the cable? Finally, in addition to the reduction of the full problem, these investigatons yield results of independent relevance.

In particular we recover a subresonant state which provides a sufficient condition for existence and uniqueness of stationary solutions u_ς of (1.1). We observe that this subresonance condition also implies the existence and uniqueness of $u = u(t)$ solving (1.1) for any time $t > 0$. There may exist stationary solutions u_ς of (1.1) which are not subresonant, but in this case we have no sufficient condition that there is a solution u of (1.1) that converges to u_ς for large times. Stationary solutions u_ς of (1.1) which are not subresonant, show an oscillatory behaviour which rather let us expect that there is no time dependent solution of (1.1) which converges to u_ς. With nonlinear boundary conditions considered, these asymptotic investigations yield new results. Moreover, we introduce the Friedrichs constant $c_\star$ induced by a physically consistent norm $\|\cdot\|_\star$ on $H^1(\Omega)$ and an associated Friedrichs-inequality $\|\cdot\|_{L^2(\Omega)} \leq c_\star \|\cdot\|_\star$.

In chapter 3 we apply our estimates to heat transfer in electric cables with an explicit geometry. The constants introduced in chapter 2 are identified with concrete physical and geometrical quantities. Here we can see, that the rather abstract conditions on the data of the initial boundary value problem in chapter 2 become plausible and provide consistent relations between the associated physical quantities. We will reveal the antagonistic behaviour of the heat transfer coefficient on the boundary Γ and the source term on the right hand side of (1.1). One at the first glance surprising result of chapter 3 is the cooling effect of insulations of electric cables, provided the heat conductivity of the insulation is large enough.

Chapter 4 deals with the reduced model (1.2). We derive a boundary integral equation for (1.2) using single and double layer potential operators. Then we propose an iterative method which solves the nonlinear boundary integral equation on $\partial\Omega_{cr}$. Again, this treatment has not only a computational motivation. It shows an interesting structure of the layer potentials in the multiply connected domain case and a damping property of harmonic functions in certain boundary geometries. The damping property means that a change of the boundary temperature changes the inner normal heat flow more than the outer normal heat flow. This can be interpreted as a plausible physical property of insulations and it is essential for the convergence of the proposed iterative procedure.

2. Asymptotic analysis of the initial boundary value problem

2.1. Asymptotics for large times

2.1.1. Setup of the initial boundary value problem

For $d \in \mathbb{N}$, $d \geq 2$ we consider a bounded domain $\Omega \subset \mathbb{R}^d$ with a Lipschitz boundary $\partial \Omega := \Gamma$. We formulate the following semilinear parabolic boundary value problem of finding the time and space dependent function $u : \Omega \times [0, \infty) \to \mathbb{R}$ such that

$$\frac{\partial u(t)}{\partial t} = \operatorname{div}(\Lambda \nabla u(t)) + \varsigma\, r(\cdot, u(t)) + f \quad \text{in } \Omega;\ t \in [0, \infty) \tag{2.1}$$

for given $\varsigma \in \mathbb{R}$.

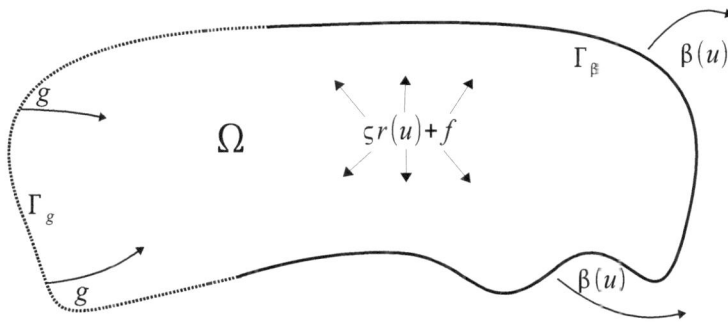

(2.1) fulfills the initial condititon $u(0) = u_{init}$ in Ω and the boundary conditions

$$-(\Lambda \nabla u(t)) n = \beta(u(t)) \quad \text{on } \Gamma_\beta \qquad (2.2)$$
$$(\Lambda \nabla u(t)) n = g \quad \text{on } \Gamma_g.$$

Here and in what follows n denotes the outer normal on Γ. In (2.2) Γ decomposes into a Neumann part Γ_g and a transmission part Γ_β, with $\Gamma_g \cap \Gamma_\beta = \emptyset$ and $\overline{\Gamma}_g \cup \overline{\Gamma}_\beta = \Gamma$. Using standard notation for Sobolev spaces, we assume

- $f \in L^2(\Omega)\,; g \in L^2(\Gamma_g)$

- $\Lambda \in L^\infty(\Omega, \mathbb{R}^{d \times d})$ is a positive definite symmetric matrix of L^∞- coefficients, i.e.

$$\exists \lambda_{min} > 0 \,:\, y\,\Lambda(x)\,y \geq \lambda_{min}\,|y|^2,\ x \in \Omega,\ y \in \mathbb{R}^d.$$

- $r : \Omega \times \mathbb{R} \to \mathbb{R}$ is a Carathéodory map which satisfies $r(\cdot, 0) = 0$ and

$$\exists L_r > 0 \,:\, |r(x, s_1) - r(x, s_2)| \leq L_r\,|s_1 - s_2|\,;\ s_1, s_2 \in \mathbb{R}. \qquad (2.3)$$

- $\beta : \mathbb{R} \to \mathbb{R}$ is continuous and satisfies the linear growth condition $\beta(s) \leq a + b\,|s|$. Moreover β satisfies the following monotonicity estimate

$$\exists c_\beta > 0 \,:\, \frac{\beta(s_1) - \beta(s_2)}{s_1 - s_2} \geq c_\beta \text{ for } s_1 \neq s_2. \qquad (2.4)$$

Remark

Lipschitz continuity of r and the growth condition on β are often too restrictive when modelling (2.1). In many applications it suffices to consider the restriction of r, β to certain compact intervals of interest and to replace r, β by suitable linear functions out of these intervals. Then the Lipschitz continuity and the growth condition are simply obtained by continuity of r, β on the compact intervals. We will make use of this remark in section 3.1.2.

Our aim is to investigate the convergence of $u(t)$ towards a stationary solution $u = u_\varsigma$ depending on $\varsigma \in \mathbb{R}$ for $t \to \infty$. Therefore we first give sufficient conditions for existence and uniqueness of a stationary solution cf (2.1). Next, using these conditions, we treat the dynamic case and its asymptotic behaviour.

2.1.2. Existence and uniqueness of a stationary solution

For given $\varsigma \in \mathbb{R}$, consider the semilinear elliptic boundary value problem of finding $u_\varsigma : \Omega \to \mathbb{R}$ that solves

$$-\operatorname{div}(\Lambda \nabla u_\varsigma) = \varsigma r(\cdot, u_\varsigma) + f \quad \text{in } \Omega \qquad (2.5)$$

subject to the boundary conditions in (2.2) and the same regularity properties of the data as listed in the section above (2.1), (2.2).

Remarks on subresonant states in elliptic equations

In physics, resonance is the response of a system that develops oscillations with large amplitudes under certain characteristic frequencies. Consider a bounded domain $\Omega \subset \mathbb{R}^2$ with a smooth boundary Γ. Then the Dirichlet eigenvalue problem $-\Delta u = \lambda u$ in Ω ; $u = 0$ on Γ describes a clamped vibrating membrane with eigenvalue λ, see e.g. [67]. The resonant frequencies ϕ_i, $i \in \mathbb{N}$ are determined by the eigenvalues λ_i, $i \in \mathbb{N}$ of $-\Delta$ under zero Dirichlet boundary conditions via $\phi_i = \sqrt{\lambda_i}\, c$, where c denotes the acoustic wave velocity of the membrane. For given $\varsigma \in \mathbb{R}$ consider now the related problem $P_{\varsigma,dir}$: For given f, find u such that there holds

$$-\Delta u = \varsigma u + f \quad \text{in } \Omega \;;\; u = 0 \text{ on } \Gamma. \qquad (2.6)$$

The problem $P_{\varsigma,dir}$ is non-resonant, if for every $f \in H^{-1}(\Omega)$ there exists a unique solution $u \in H^1(\Omega)$; otherwise it is resonant [24]. The Lax-Milgram Theorem and Friedrichs' inequality tell us that there is a nonresonant state of (2.5) if $|\varsigma| < \lambda_1$; where λ_1 is the principal eigenvalue of $(-\Delta, H_0^1(\Omega))$. We call this state subresonant. As is well known, a variational form of λ_1 is given by the Rayleigh-quotient $\lambda_1 = \min\limits_{u \in H_0^1(\Omega) \setminus \{0\}} \frac{\|\nabla u\|_{L^2(\Omega)}^2}{\|u\|_{L^2(\Omega)}^2}$. Alternatively, subresonance can be described by the associated Friedrichs constant

$$c_F(\Omega) = \sup_{u \in H_0^1(\Omega) \setminus \{0\}} \frac{\|u\|_{L^2(\Omega)}}{\|\nabla u\|_{L^2(\Omega)}} = \frac{1}{\sqrt{\lambda_1}},$$

i.e. the optimal constant in Friedrichs' inequality $\|u\|_{L^2(\Omega)} \leq c_F(\Omega) \|\nabla u\|_{L^2(\Omega)}$ for $u \in H_0^1(\Omega)$.

Here we note that the solvability of the linear problem in (2.6) is obvious for $\varsigma \leq 0$, including $|\varsigma| \geq \lambda_1$. Thus resonance can only occur for $\varsigma \geq \lambda_1$. If semilinear or quasilinear elliptic Dirichlet problems $-Lu = \varsigma\, r(u) + h$ in Ω; $u = 0$ on Γ are considered, the solvability problem remains relevant for $\varsigma \leq 0$. Existence results can be given, see e.g. [49], [50] [57], [58] for classical treatments and [23], [51], [9] for more recent papers.

Construction of a physically consistent norm on $H^1(\Omega)$

In the sequel, we introduce a Friedrichs constant $c_\star$ induced by a physically consistent norm $\|\cdot\|_\star$ on $H^1(\Omega)$. Then we give sufficient conditions for subresonance in (2.5) via $c_\star$ and provide an explicit bound on u_ς in the norm $\|\cdot\|_\star$. Let $x \in \Omega$ denote a space variable measured in L and v an arbitrary physical quantity measured in V. Thus we obtain $V^2 L^{d-2}$ as a unit for $\|\nabla v\|_{L^2(\Omega)}^2$ and $V^2 L^{d-1}$ as a unit for $\|v\|_{L^2(\Gamma_\beta)}^2$. (2.2) implies that the quotient $\frac{c_\beta}{\lambda_{min}}$ is measured by L^{-1}. Respecting the question of units, we equip the Sobolev space $H^1(\Omega)$ with the physically consistent seminorm

$$\|v\|_\star^2 := \|\nabla v\|_{L^2(\Omega)}^2 + \frac{c_\beta}{\lambda_{min}} \|v\|_{L^2(\Gamma_\beta)}^2 \,.$$

If the $(d-1)$-dimensional Hausdorff-measure of Γ_β ($|\Gamma_\beta|$ for short) is positive, $\|\cdot\|_\star$ is equivalent to the canonical H_1-norm denoted by $\|\cdot\|$. In fact we have the following Lemma.

Lemma 2.1

Let $|\Gamma_\beta| > 0$. Then there exist $c_1, c_2 > 0$ not depending on v, such that $c_1 \|v\|_\star \leq \|v\| \leq c_2 \|v\|_\star$, $\forall v \in H^1(\Omega)$.

Proof

(i) Let $C_{\Gamma_\beta} := \|\tau\|_{tr} = \sup\limits_{\|v\|\leq 1} \|\tau(v)\|_{L^2(\Gamma_\beta)}$ denote the norm of the trace map $\tau : H^1(\Omega) \to L^2(\Gamma_\beta)$. Then the first inequality follows from

$$\|v\|_\star^2 \leq \|\nabla v\|_{L^2(\Omega)}^2 + \frac{c_\beta\, C_{\Gamma_\beta}^2}{\lambda_{min}} \|v\|^2 \leq \left(1 + \frac{c_\beta\, C_{\Gamma_\beta}^2}{\lambda_{min}}\right) \|v\|^2 ,$$

Thus we have $c_1 = \left(1 + \frac{c_\beta\, C_{\Gamma_\beta}^2}{\lambda_{min}}\right)^{-1/2}$.

(ii) Define the Friedrichs constant $c_\star := \sup\limits_{v \in H^1(\Omega) \setminus \{0\}} \left(\|v\|_{L^2(\Omega)} / \|v\|_\star\right)$. Then there holds $c_2 = \sqrt{1 + c_\star^2}$. □

Variational formulation of (2.5)

For $u, v \in H^1(\Omega)$ we define the nonlinear operator A and the linear form b by

$$\langle Au, v \rangle := \int_\Omega \nabla u \wedge \nabla v \, dx + \int_{\Gamma_\beta} \beta(u)\, v \, d\sigma - \int_\Omega \varsigma r(x, u)\, v \, dx \quad (2.7)$$

$$\langle b, v \rangle := \int_\Omega f v \, dx + \int_{\Gamma_g} g v \, d\sigma$$

where $\langle \cdot, \cdot \rangle$ denotes the duality pairing between $(H^1(\Omega))_\star^*$ and $H^1(\Omega)$. We show that the growth condition on β and the Lipschitz-condition on r imply the mapping property $A : H^1(\Omega) \to (H^1(\Omega))^*$.

Lemma 2.2

Let A denote the operator defined in (2.7). Then for every $u \in H^1(\Omega)$ there holds $Au \in (H^1(\Omega))^*$.

Proof

$\langle Au, v \rangle$ is linear in v, hence it suffices to show that $\langle Au, \cdot \rangle$ is bounded.

$\Lambda \in L^\infty(\Omega, \mathbb{R}^{d \times d})$ implies $\exists \lambda_{max} < \infty$: $\operatorname*{ess\,sup}_{x \in \Omega} (y_1 \Lambda(x) y_2) \leq \lambda_{max} |y_1| |y_2|$, hence we have

$$|\langle Au, v \rangle| \leq \lambda_{max} \|\nabla u\|_{L^2(\Omega)} \|\nabla v\|_{L^2(\Omega)}$$
$$+ \|a + b\,|u|\|_{L^2(\Gamma_\beta)} \|v\|_{L^2(\Gamma_\beta)} + |\varsigma|\, L_r \|u\|_{L^2(\Omega)} \|v\|_{L^2(\Omega)} \,.$$

I.e. $\exists C < \infty$ not depending on v such that $|\langle Au, v \rangle| \leq C \|v\|_\star$. $\square$

Thus the variational form of (2.5) reads as

$$\langle Au_\varsigma, v \rangle = \langle b, v \rangle \quad \forall v \in H^1(\Omega). \tag{2.8}$$

Theorem 2.1
Let $|\varsigma| < \frac{\lambda_{min}}{L_r c_\star^2}$. Then, for all $f \in L^2(\Omega)$, $g \in L^2(\Gamma_g)$ there exists a unique solution $u_\varsigma \in H^1(\Omega)$ of (2.8) which is bounded by

$$(\lambda_{min} - L_r c_\star^2 |\varsigma|) \|u_\varsigma\|_\star \leq c_\star \|f\|_{L^2(\Omega)} + c_{L^2} \|g\|_{L^2(\Gamma_g)} + \sqrt{\frac{|\Gamma_\beta| \lambda_{min}}{c_\beta}} |\beta(0)| \,.$$

$c_{L^2} := \|\tau\|_{tr} = \sup_{\|v\|_\star \leq 1} \|\tau(v)\|_{L^2(\Gamma_g)}$ denotes the norm of the trace map $\tau : H^1(\Omega) \to L^2(\Gamma_g)$. We will give an explicit estimate of c_{L^2} in section 2.4.3.

Proof of Theorem 2.1
(i) existence and uniqueness
We consider the variational formulation in (2.8). The monotonicity condition in (2.4) and the assumption on ς above implies the strong monotonicity of the operator $A : H^1(\Omega) \to (H^1(\Omega))^*$.

$$\langle Au - Av, u - v \rangle \geq \lambda_{min} \|\nabla(u-v)\|^2_{L^2(\Omega)} + \langle \beta(u) - \beta(v), u - v \rangle_{L^2(\Gamma_\beta)}$$
$$- |\varsigma| L_r \|u-v\|^2_{L^2(\Omega)}$$
$$\geq \lambda_{min} \|u-v\|^2_\star - |\varsigma| L_r \|u-v\|^2_{L^2(\Omega)}$$
$$\geq (\lambda_{min} - L_r c_\star^2 |\varsigma|) \|u-v\|^2_\star$$

The hemicontinuity of A i.e. the continuity of $s \mapsto \langle A(u+sv), w \rangle$; $s \in [0,1]$ for $u, v, w \in H^1(\Omega)$ follows from the continuity of r and β. Thus existence and uniqueness follow by the Theorem of Browder and Minty for monotone operators, (A.2).

(ii) boundedness
We have $\langle b, u_\varsigma \rangle \leq \left(c_\star \|f\|_{L^2(\Omega)} + c_{L^2} \|g\|_{L^2(\Gamma_g)} \right) \|u_\varsigma\|_\star$
and on the other hand

$$\langle A u_\varsigma, u_\varsigma \rangle \geq (\lambda_{min} - L_r c_\star^2 |\varsigma|) \|u_\varsigma\|_\star^2 + \langle \beta(0), u_\varsigma \rangle_{L^2(\Gamma_\beta)}$$
$$\geq (\lambda_{min} - L_r c_\star^2 |\varsigma|) \|u_\varsigma\|_\star^2 + \sqrt{\frac{|\Gamma_\beta| \lambda_{min}}{c_\beta}} |\beta(0)| \|u_\varsigma\|_\star$$

which implies the result. □

Damping effect for negative values of ς

If r is monotonically increasing and $\varsigma < 0$, then the bound in Theorem 2.1 holds with

$$\lambda_{min} \|u_\varsigma\|_\star \leq c_\star \|f\|_{L^2(\Omega)} + c_{L^2} \|g\|_{L^2(\Gamma_g)} + \sqrt{\frac{|\Gamma_\beta| \lambda_{min}}{c_\beta}} |\beta(0)| \quad (2.9)$$

for arbitrarily large values of $|\varsigma|$. This is due to the damping effect of $\varsigma r(\cdot, u)$ in this case. The according estimate is easily seen from the proof of Theorem 2.1. It also includes the classical result that solutions of linear Neumann boundary value problems $-\Delta u = cu + f$ in Ω ; $\frac{\partial u}{\partial n} = 0$ on Γ exist uniquely for $c < 0$. The solution is bounded by $\|u\|_{H^1(\Omega)} \leq \|f\|_{L^2(\Omega)}$. An explicit monotonicity condition on r of the form
$\exists c_r > 0 : \inf_{x,y \in \Omega} \left(\frac{r(x,s_1) - r(x,s_2)}{s_1 - s_2} \right) \geq c_r$ for $s_1 \neq s_2$ and $\varsigma < 0$ cannot improve the estimate in (2.9). This is due to the irreversibility of Friedrichs inequality $\|\cdot\|_{L^2(\Omega)} \leq c_\star \|\cdot\|_\star$ used in the proof of Theorem 2.1. This obstacle vanishes if we consider temperature profiles constant in space as in section 2.1.4.

2.1.3. Treatment of the dynamical problem

Existence and uniqueness of the dynamical solution

Now we consider the dynamical problem in (2.1), using the strongly monotone operator $A : H^1(\Omega) \to (H^1(\Omega))^*$ and the linear form $b \in (H^1(\Omega))^*$ defined in (2.7). Thus the variational form of (2.1) reads as

$$\left\langle \frac{\partial u(t)}{\partial t}, v \right\rangle + \langle A u(t), v \rangle = \langle b, v \rangle \quad \forall v \in H^1(\Omega) \tag{2.10}$$

$$u(0) = u_{init} \in H^1(\Omega).$$

Following an approach of H. Brézis ([12],chap.III) we give (2.10) a rigorous treatment, considering the evolution $[0, \infty) \ni t \mapsto u(t) \in H^1(\Omega)$ as an element of the Bochner space
$L^1([0, \infty), H^1(\Omega)) := \{u : [0, \infty) \to H^1(\Omega) \, ; \, \int_0^\infty \|u(t)\|_\star \, dt < \infty\}$. Thus we identify the time derivative $\frac{\partial u}{\partial t}$ as an element of $L^\infty((0, \infty), (H^1(\Omega))^*)$ in the sense of distributions. In particular, we have $\frac{\partial u(t)}{\partial t} \in (H^1(\Omega))^*$ and the duality pairing $\left\langle \frac{\partial u(t)}{\partial t}, v \right\rangle$ in (2.10) is well defined. See also [7], [61] or [64] for further investigations on nonlinear evolution equations.

Theorem 2.2 (Existence and uniqueness of $u(t)$)
Let $A : H^1(\Omega) \to (H^1(\Omega))^*$ be strongly monotone and $b \in (H^1(\Omega))^*$. Then there exists a Lipschitz-continuous and unique evolution $[0, \infty) \ni t \mapsto u(t) \in H^1(\Omega)$ satisfying (2.10).

Proof
Observe that the operator $B(u) := A(u) - b$ is still strongly monotone and thus maximally monotone. Then (2.10) reads as $\frac{\partial u}{\partial t} + B(u) = 0$ in $(H^1(\Omega))^*$ and the assertions follow by Theorem 3.1 in [12]. □

Convergence to the stationary solution

Using the subresonance condition $|\varsigma| < \frac{\lambda_{min}}{L_r c_\star^2}$ of Theorem 2.1 we obtain strong monotonicity of $A : H^1(\Omega) \to (H^1(\Omega))^*$. Thus we have the existence of

$(u(t))_{t\in[0,\infty)} \subset H^1(\Omega)$ solving the initial boundary value problem (2.1).

Proposition 2.1
Let $|\varsigma| < \frac{\lambda_{min}}{L_r c_\star^2}$ and let $(u(t))_{t\in[0,\infty)} \subset H^1(\Omega)$, $u_\varsigma \in H^1(\Omega)$ denote the solutions of (2.1), (2.5) respectively. Then there holds

$$\|u(t) - u_\varsigma\|_{L^2(\Omega)} \leq e^{-\phi t} \|u_{init} - u_\varsigma\|_{L^2(\Omega)} \quad \text{where } \phi := \frac{\lambda_{min}}{c_\star^2} - L_r |\varsigma| . \quad (2.11)$$

Proof
Note that the stationary solution u_ς trivially satisfies the equation $\frac{\partial u_\varsigma}{\partial t} + A(u_\varsigma) = b$ in $(H^1(\Omega))^*$. Hence the chain rule and the strong monotonicity of A yield

$$\begin{aligned} \frac{1}{2}\frac{\partial}{\partial t}\left(\|u(t) - u_\varsigma\|_{L^2(\Omega)}^2\right) &= \left\langle \frac{\partial u(t)}{\partial t} - \frac{\partial u_\varsigma}{\partial t}, u(t) - u_\varsigma \right\rangle \\ &= \langle A(u_\varsigma) - A(u(t)), u(t) - u_\varsigma \rangle \\ &\leq -\left(\lambda_{min} - L_r c_\star^2 |\varsigma|\right) \|u(t) - u_\varsigma\|_\star^2 \\ &\leq -\left(\frac{\lambda_{min}}{c_\star^2} - L_r |\varsigma|\right) \|u(t) - u_\varsigma\|_{L^2(\Omega)}^2 \end{aligned}$$

Thus the function $y(t) := \|u(t) - u_\varsigma\|_{L^2(\Omega)}^2$ satisfies the inequality $\dot{y}(t) \leq -2\phi y(t)$, $t \in [0,\infty)$. Gronwall's inequality, (A.4) provides $y(t) \leq y(0) e^{-2\phi t}$ which implies (2.11). □

Remarks
(i) The sufficient condition $|\varsigma| < \frac{\lambda_{min}}{L_r c_\star^2}$ for the existence of u_ς in Theorem 2.1 also implies the existence of the whole evolution $(u(t))_{t\in[0,\infty)}$ solving (2.1). Moreover $(u(t))$ converges exponentially to u_ς by Proposition 2.1. In chapter 3, we will apply the estimate (2.11) to heat transfer in electric cables.

(ii) If r is monotonically increasing and $\varsigma \leq 0$, then the result of Theorem 2.2 holds for arbitrarily large $|\varsigma|$ with $\phi = \frac{\lambda_{min}}{c_\star^2}$. The respective estimate follows directly from the proof of Theorem 2.2 and the considerations in section 2.1.2.

(iii) The estimate (2.11) is given in the L^2- Norm since $L^2(\Omega)$ is the appropriate interpolating Hilbert space between $H^1(\Omega)$ and its dual via the Gelfand triple $H^1(\Omega) \subset L^2(\Omega) \subset (H^1(\Omega))^*$.

Interpolation between u_{init} and u_ς

If the initial datum u_{init} and the stationary solution u_ς of (2.5) are known, we can interpolate by $\tilde{u}(t) := e^{-\phi t} u_{init} + (1 - e^{-\phi t}) u_\varsigma$. By Proposition 2.1, it approximates the original evolution $u = u(t)$ of (2.1) with the following error bound for large times.

$$\|u(t) - \tilde{u}(t)\|_{L^2(\Omega)} \leq 2 e^{-\phi t} \|u_{init} - u_\varsigma\|_{L^2(\Omega)}$$

For small times we note that $u(0) = \tilde{u}(0)$. Moreover we can compare the time derivatives of u and $\tilde{u}$ deriving the following result.

Lemma 2.3
Let $|\varsigma| < \frac{\lambda_{min}}{L_r c_*^2}$ and let $(u(t))_{t\in[0,\infty)} \subset H^1(\Omega)$, $(\tilde{u}(t))_{t\in[0,\infty)} \subset H^1(\Omega)$ denote the solution of (2.1) and the approximating interpolation respectively. Moreover suppose $\frac{\partial u(0)}{\partial t} \in L^2(\Omega)$. Then there holds

$$\left\|\frac{\partial u(0)}{\partial t}\right\|_{L^2(\Omega)} \geq \left\|\frac{\partial \tilde{u}(0)}{\partial t}\right\|_{L^2(\Omega)}.$$

Proof
By the definition of $\tilde{u}$ we have $\frac{\partial \tilde{u}(0)}{\partial t} = \phi(u_\varsigma - u_{init}) \in H^1(\Omega) \subset L^2(\Omega)$. On the other hand - using (2.10) - there holds

$$\left\langle \frac{\partial u(0)}{\partial t}, v \right\rangle + \langle A u_{init} - b, v \rangle = 0 \ \forall v \in H^1(\Omega)$$
$$\iff \left\langle \frac{\partial u(0)}{\partial t}, v \right\rangle + \langle A u_{init} - A u_\varsigma, v \rangle = 0 \ \forall v \in H^1(\Omega).$$

Setting $v = u_\varsigma - u_{init}$, using the monotonicity of A and the definition of ϕ we get $\left\langle \frac{\partial u(0)}{\partial t}, u_\varsigma - u_{init} \right\rangle \geq c_\star^2 \phi \, \|u_\varsigma - u_{init}\|_\star^2$. The Cauchy-Schwarz inequality and $\|\cdot\|_{L^2(\Omega)} \leq c_\star \|\cdot\|_\star$ imply

$$\left\| \frac{\partial u(0)}{\partial t} \right\|_{L^2(\Omega)} \geq \phi \, \|u_\varsigma - u_{init}\|_{L^2(\Omega)}$$

and thus the assertion. □

The following diagramm illustrates qualitatively the temperature evolution u at a point $x \in \Omega$ and the associated interpolating approximation $\tilde{u}$.

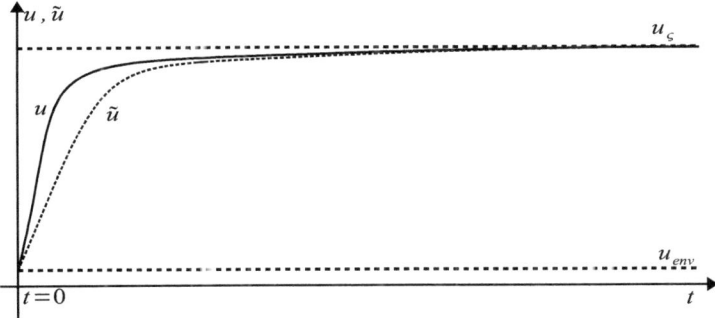

This shows that $\tilde{u}$ and u have the same asymptotic behaviour; and - by Lemma 2.3 - $\tilde{u}$ is a lower bound for u in a neighbourhood of $t = 0$.

2.1.4. Analysis of (2.1) for constant temperature profiles

Suppose $u = u(t)$ describes an evolution of a temperature profile in Ω. Moreover, suppose in (2.1) that we have a homogeneous Neumann datum $g = 0$ on Γ_g and an autonomous resonance map $r = r(u)$. On the other hand assume that λ_{min} is comparatively large; e.g. the heat conductivity of a metallic conductor. Thus the associated temperature profile evolution $(u(t))_{t \in [0,\infty)}$ is almost constant in space.

In this case it makes sense to approximate the evolution $u = u(t)$ by an implicitly defined energy conserving mean value $(u^{mv}(t))_{t\in[0,\infty)} \subset H^1(\Omega)$ which is constant in space. The mean value evolution u^{mv} is defined by the variational formulation of (2.1) in (2.10), i.e.

$$\left\langle \frac{\partial u^{mv}(t)}{\partial t}, v \right\rangle + \langle A u^{mv}(t), v \rangle = \langle b, v \rangle \quad \forall v \in H^1(\Omega); \; t \in (0, \infty)$$
$$u^{mv}(0) = u^{mv}_{init} \in \mathbb{R}$$

where $\langle A u^{mv}(t), v \rangle = \int_{\Gamma_\beta} \beta(u^{mv}(t))\, v\, d\sigma - \int_\Omega \varsigma\, r(u^{mv}(t))\, v\, dx$
and $\langle b, v \rangle = \int_\Omega f v\, dx$. Setting $v = 1$ we obtain the ordinary differential equation

$$\dot{u}^{mv} = \varsigma\, r(u^{mv}) + \frac{1}{|\Omega|} \int_\Omega f\, dx - \frac{|\Gamma_\beta|}{|\Omega|} \beta(u^{mv}) \quad \text{in } (0, \infty) \quad (2.12)$$
$$u^{mv}(0) = u^{mv}_{init}.$$

Proposition 2.2 (Existence and uniqueness of u^{mv})
Let $r \in C(\mathbb{R})$ and $\beta \in C(\mathbb{R})$ fulfill the Lipschitz and the monotonicity condition in (2.3) and (2.4). Then there exists a unique solution $u^{mv} \in C^1((0,\infty))$ of (2.12) for every $\varsigma \in \mathbb{R}$.

Proof
We show that the right hand side $F : \mathbb{R} \to \mathbb{R}$ of (2.12), given by

$$F(s) := \varsigma\, r(s) + \frac{1}{|\Omega|} \int_\Omega f\, dx - \frac{|\Gamma_\beta|}{|\Omega|} \beta(s)$$

satisfies a global Lipschitz condition on $\mathbb{R}$. Using (2.3) and (2.4), there holds

$$F(u) - F(v) = \varsigma\, (r(u) - r(v)) - \frac{|\Gamma_\beta|}{|\Omega|}(\beta(u) - \beta(v))$$
$$\leq |\varsigma|\, L_r\, |u - v| - \frac{|\Gamma_\beta| c_\beta}{|\Omega|}(u - v) \leq \left(|\varsigma|\, L_r + \frac{|\Gamma_\beta| c_\beta}{|\Omega|} \right) |u - v|.$$

On the other hand we have

$$F(v) - F(u) = \varsigma\left(r(v) - r(u)\right) - \frac{|\Gamma_\beta|}{|\Omega|}(\beta(v) - \beta(u))$$
$$\leq |\varsigma| L_r |u - v| - \frac{|\Gamma_\beta| c_\beta}{|\Omega|}(v - u) \leq \left(|\varsigma| L_r + \frac{|\Gamma_\beta| c_\beta}{|\Omega|}\right)|u - v|.$$

which implies $|F(u) - F(v)| \leq \left(|\varsigma| L_r + \frac{|\Gamma_\beta| c_\beta}{|\Omega|}\right)|u - v|$ for arbitrary $u, v \in \mathbb{R}$. Thus the assertion of Proposition (2.2) follows by the global version of the Picard-Lindelöf Theorem. $\square$

Existence and uniqueness of a stationary solution

By Proposition 2.2 there exists an evolution u^{mv} in $(0, \infty)$ for arbitrary $\varsigma \in \mathbb{R}$. Nevertheless, this evolution can grow unboundedly and no stationary solution $u_{st} \in \mathbb{R}$ of (2.12) that satifies

$$\varsigma r(u_{st}) + \frac{1}{|\Omega|}\int_\Omega f \, dx = \frac{|\Gamma_\beta|}{|\Omega|}\beta(u_{st}) \tag{2.13}$$

exists. The following Corollary gives a sufficient condition for existence and uniqueness of a stationary solution.

Corollary 2.1

Suppose that the conditions of Proposition 2.2 and the relation $|\varsigma| L_r < \frac{\Gamma_\beta}{|\Omega|} c_\beta$ hold. Then there exists a unique solution $u_{st} \in \mathbb{R}$ of (2.13).

Proof

We show that the continuous map $h : \mathbb{R} \to \mathbb{R}$ with $h(s) := \varsigma r(s) - \frac{|\Gamma_\beta|}{|\Omega|}\beta(s)$ is strictly monotonically decreasing. Suppose $s_1 < s_2$, then

$$h(s_2) - h(s_1) = \varsigma\left(r(s_2) - r(s_1)\right) - \frac{|\Gamma_\beta|}{|\Omega|}(\beta(s_2) - \beta(s_1))$$
$$\leq \left(|\varsigma| L_r - \frac{c_\beta |\Gamma_\beta|}{|\Omega|}\right)|s_2 - s_1| < 0$$

Thus the equation $h(s) - \frac{1}{|\Omega|} \int_\Omega f \, dx = 0$ has a unique solution in $\mathbb{R}$. $\square$

Remark
In applications to heat transfer in uninsulated cables we have a unique stationary solution in (2.12) even for $|\varsigma| L_r > \frac{\Gamma_\beta}{|\Omega|} c_\beta$. This is due to the specific structure of r and β in that case. We will discuss this in chapter 3.

Asymptotic behaviour of u^{mv} for $t \to \infty$

If there exists a stationary state u_{st} of (2.12), we can investigate the convergence of $u^{mv} \xrightarrow[t \to \infty]{} u_{st}$ in $\mathbb{R}$.

Corollary 2.2
Suppose that the conditions of Corollary 2.1 hold. Let u^{mv} and u_{st} denote the solutions of (2.12) and (2.13) respectively. Then

$$|u^{mv}(t) - u_{st}| \leq e^{-\phi^{mv} t} |u^{mv}_{init} - u_{st}| \; ; \; \phi^{mv} := \frac{|\Gamma_\beta| c_\beta}{|\Omega|} - |\varsigma| L_r. \qquad (2.14)$$

Proof
Observe that u_{st} satisfies (2.12). Thus we have

$$\dot{u}^{mv}(t) - \dot{u}_{st} = \varsigma \left(r(u^{mv}) - r(u_{st}) \right) - \frac{|\Gamma_\beta|}{|\Omega|} \left(\beta(u^{mv}) - \beta(u_{st}) \right)$$

$$\leq \left(|\varsigma| L_r - \frac{|\Gamma_\beta| c_\beta}{|\Omega|} \right) |u^{mv}(t) - u_{st}| = -\phi^{mv} |u^{mv}(t) - u_{st}|.$$

The same estimate holds for $\dot{u}_{st} - \dot{u}^{mv}(t)$. Hence $y(t) := |u^{mv}(t) - u_{st}|$ satisfies $\dot{y}(t) \leq -\phi^{mv} y(t)$ and Gronwall's inequality implies $y(t) \leq y(0) e^{-\phi^{mv} t}$. $\square$

Improved convergence for monotonically increasing r and negative ς
Suppose that r fulfills the monotonicity condition

$$\exists c_r > 0 : \left(\frac{r(s_1) - r(s_2)}{s_1 - s_2} \right) \geq c_r \text{ for } s_1 \neq s_2. \qquad (2.15)$$

Then a negative ς extends the existence range (subresonant state) of (2.13). I.e. if $\frac{|\Gamma_\beta| c_\beta}{|\Omega|} - \varsigma c_r > 0$ then there exists a unique solution u_{st} of (2.13). Moreover the rate of convergence of u^{mv} towards u_{st} in (2.14) is improved by $\phi^{mv} := \frac{|\Gamma_\beta| c_\beta}{|\Omega|} - \varsigma c_r$. This is easily seen by an application of the monotonicity property on r (2.15) in the proof of Corollary 2.2.

Computation of u^{mv} for finite times

Assuming the conditions of Proposition 2.2 we have a unique solution $u^{mv} = u^{mv}(t)$ of (2.12) in $[0, \infty)$. The aim of this paragraph is to provide methods for the computation of u^{mv} in a finite time interval $[0, t_{max}]$. Namely we use the Picard iteration and the explicit Euler scheme.

Picard iteration

The following Corollary provides an iterative approximating sequence $(z_n)_{n \in \mathbb{N}} \subset C([0, t_{max}])$ to the solution of (2.12).

Corollary 2.3
Let $z_1 \in C([0, t_{max}])$ denote an arbitrary initial function Then $(z_n)_{n \in \mathbb{N}} \subset C([0, t_{max}])$ defined iteratively by

$$z_{n+1}(t) = u^{mv}_{init} + \int_0^t \left(\varsigma r(z_n(s)) - \frac{|\Gamma_\beta|}{|\Omega|} \beta(z_n(s)) + \frac{1}{|\Omega|} \int_\Omega f \, dx \right) ds, \quad t \in [0, t_{max}]$$

converges uniformly in $C([0, t_{max}])$ to the solution of (2.12).

The proof makes use of Banach's Fixed Point Theorem in $C([0, t_{max}])$. For details and associated error estimates we refer to [20].

Euler Scheme
To illustrate the scheme we divide the interval $[0, t_{max}]$ in n subintervals of length $\delta = \frac{t_{max}}{n}$ and denote the corresponding nodal points with $t_i = (i-1)\delta$, $i = 1, \ldots n+1$. We approximate the derivative in (2.12) with a forward dif-

ference scheme:

$$\frac{u_{i+1}^{mv} - u_i^{mv}}{\delta} = \varsigma\, r(u_i^{mv}) + \frac{1}{|\Omega|}\int_\Omega f\,dx - \frac{|\Gamma_\beta|}{|\Omega|}\beta(u_i^{mv}) =: F(u_i^{mv})\,;\ i=1,\ldots,n$$

which yields the explicit Euler algorithm for (2.12)

$$u_1^{mv} = u_{init}^{mv};\ u_{i+1}^{mv} = u_i^{mv} + \delta\, F(u_i^{mv})\,;\ i=1,\ldots,n\,. \tag{2.16}$$

Let $y_n \in C([0, t_{max}])$ denote the associated linear interpolation of the nodal points $(t_i, u_i^{mv})_{i=1}^{n+1}$ in $C([0, t_{max}])$. Suppose now that the data r and β in (2.12) are sufficiently smooth; such that the solution of (2.12) is twice continuously differentiable. Thus we obtain

Corollary 2.4
Let r, $\beta \in C^1(\mathbb{R})$ fulfill the assumpions of Proposition 2.2. Then the linear interpolation $(y_n)_{n\in\mathbb{N}} \subset C([0, t_{max}])$ defined by the explicit Euler scheme in (2.16) converges uniformly in $C([0, t_{max}])$ to the solution of (2.12)

We refer to [38] for the proof and the respective error bounds.

Exponential growth estimate of u^{mv} for superlinear r and sublinear β

For $\varsigma \geq \frac{\lambda_{min}}{L_r c_*^2}$ the existence of a stationary solution of (2.5) is not ensured and so the asymptotic behavior of solutions of (2.1) is unclear. Nevertheless, for a sufficiently large ς and suitable conditions on r and β it is possible to establish an exponential growth estimate for solutions of (2.1). As an instructive case we consider the homogeneous initial boundary value problem

$$\begin{aligned}\frac{\partial u(t)}{\partial t} &= \operatorname{div}(\Lambda\, \nabla u(t)) + \varsigma\, r(\cdot, u(t)) \text{ in } \Omega;\ t \in (0, t_{max}) \\ -(\Lambda\, \nabla u(t))\,n &= \beta(u(t)) \quad \text{on } \Gamma_\beta\,;\quad -(\Lambda\,\nabla u(t))\,n = 0 \quad \text{on } \Gamma_g\end{aligned} \tag{2.17}$$

subject to the initial condititon $u(0) = u_{init} \in H^1(\Omega)$. Assume that there exists an evolution $[0, t_{max}] \ni t \mapsto u(t) \in H^1(\Omega)$ satisfying (2.17) in the weak sense, i.e.

$$\left\langle \frac{\partial u(t)}{\partial t}, v \right\rangle + \langle A u(t), v \rangle = 0 \ \forall v \in H^1(\Omega); \ t \in (0, t_{max})$$
$$u(0) = u_{init} \in H^1(\Omega)$$

where $A : H^1(\Omega) \to (H^1(\Omega))^*$ is defined as in the proof of Theorem 2.1.
In addition to this evolution we consider again the implicitly defined energy conservating mean value $(u^{mv}(t))_{t \in [0, t_{max}]} \subset H^1(\Omega)$ which is constant in space via

$$\left\langle \frac{\partial u^{mv}(t)}{\partial t}, v \right\rangle = -\langle A u^{mv}(t), v \rangle \ \forall v \in H^1(\Omega) \quad (2.18)$$
$$u^{mv}(0) = u^{mv}_{init} \in \mathbb{R}$$

where $\langle A u^{mv}(t), v \rangle = \int_{\Gamma_\beta} \beta(u^{mv}(t)) v \, d\sigma - \int_{\Omega} \varsigma r(x, u^{mv}(t)) v \, dx$.
In the following Proposition we show: If $\varsigma > 0$ is chosen large enough then - for every $u^{mv}_{init} \in \mathbb{R} \setminus \{0\}$ - $u^{mv}(t)$ increases exponentially in time.
For an explicit treatment we require a sublinear growth condition on the boundary transfer map $\beta \in C(\mathbb{R})$

$$\exists L_\beta > 0 : |\beta(s)| \leq L_\beta |s| \ \text{for} \ s \in \mathbb{R}$$

and a superlinear growth condition on the resonance map $r \in C(\Omega \times \mathbb{R})$

$$\exists r_{min} > 0 : \inf_{x \in \Omega} r(x, s) \geq r_{min} |s|, \ s \in \mathbb{R}.$$

Observe that these 'intensifying' requirements are inverse to the 'damping' requirements for the subresonant case in the previous paragraphs.
Setting $v = 1$ in (2.18) we obtain the ordinary differential equation

$$\dot{u}^{mv} = \frac{\varsigma}{|\Omega|} \int_\Omega r(x, u^{mv}) \, dx - \frac{|\Gamma_\beta|}{|\Omega|} \beta(u^{mv}) \ \text{in} \ (0, t_{max}) \quad (2.19)$$
$$u^{mv}(0) = u^{mv}_{init}$$

The existence of a solution to (2.19) is guaranteed by Peano's theorem due to the continuity of the right hand side. Since $r \in C(\Omega \times \mathbb{R})$ has a superlinear growth, this result holds in possibly arbitrarily small interval $[0, \delta] \subset [0, t_{max}]$ only. For the following we assume the existence of a solution of (2.19) in the whole interval $[0, t_{max}]$ and formulate

Proposition 2.3
Let $(u^{mv}(t))_{t \in [0, t_{max}]} \subset \mathbb{R}$ denote a solution of (2.19) and let $\varsigma \geq \frac{|\Gamma_\beta| L_\beta}{|\Omega| r_{min}}$ then there holds

$$|u^{mv}(t)| \geq |u^{mv}_{init}| e^{\phi_{res} t} \quad \text{where} \quad \phi_{res} := \varsigma \, r_{min} - \frac{|\Gamma_\beta| L_\beta}{|\Omega|}.$$

Proof
We set $u^{mv} = v$ for short. (2.19) and the growth conditions on β and r imply $\frac{1}{2} \frac{d}{dt}(v^2) = \dot{v} v \geq \varsigma \, r_{min} v^2 - \frac{|\Gamma_\beta|}{|\Omega|} L_\beta v^2$. This reads as $\frac{d}{dt}(v^2) \geq 2 \phi_{res} v^2$. Now $y(t) = v^2(t)$ and an integration of the inequality above yields $y(t) \geq y(0) e^{2 \phi_{res} t}$, i.e. the assertion. $\square$

Remarks
(i) The monotonicity condition on β and the Lipschitz condition on r are no longer needed in the treatment above. Nevertheless we require an existence argument for (2.17), i.e. for parabolic equations with superlinear growth on the right hand side. See e.g. [69] for existence and uniqueness/non-uniqueness results.

(ii) We will apply this exponential growth to heat transfer in electric cables in chapter 2.

2.2. Approximation of subresonant solutions

In the next reduction step of (2.1) we neglect the nonlinear term $\varsigma \, r(\,\cdot\,, u)$. For given $\varsigma \in \mathbb{R}$ we consider the problem P_ς in (2.5) and study the approximation of P_ς by P_0, i.e by

$$-\operatorname{div}(\Lambda \nabla u_0) = f \quad \text{in } \Omega \qquad (2.20)$$
$$-(\Lambda \nabla u_0) n = \beta(u_0) \quad \text{on } \Gamma_\beta ; \quad (\Lambda \nabla u_0) n = g \quad \text{on } \Gamma_g.$$

providing an estimate for the resulting error in the norm $\|\cdot\|_\star$.

2.2.1. Sensitivity Results

Proposition 2.4
Let u_ς, u_0 denote the solutions of the boundary value problems P_ς, P_0 respectively. Then there holds $\limsup_{\varsigma \to 0} \frac{\|u_\varsigma - u\|_\star}{|\varsigma|} < \infty$

Proof
Consider the difference in the variational equations of P_ς and P_0 i.e.

$$\langle Au_\rho - Au_0, v \rangle = 0 \quad \forall v \in H^1(\Omega).$$

This reads as

$$\int_\Omega (\nabla u_\varsigma - \nabla u_0) \Lambda \nabla v \, dx + \int_\Gamma (\beta(u_\varsigma) - \beta(u_0)) v \, d\sigma_x = \varsigma \int_\Omega r(x, u_\varsigma) v \, dx.$$

Set $v = u_\varsigma - u_0$ and we obtain $\lambda_{min} \|u_\varsigma - u_0\|_\star^2 \leq \varsigma \int_\Omega \alpha(x, u_\varsigma)(u_\varsigma - u_0) \, dx$.
Lipschitz-continuity of r and the Cauchy-Schwarz inequality imply

$$\lambda_{min} \|u_\varsigma - u_0\|_\star^2 \leq |\varsigma| L_r \|u_\varsigma\|_{L^2(\Omega)} \|u_\varsigma - u_0\|_{L^2(\Omega)}.$$

This gives $\lambda_{min} \|u_\varsigma - u_0\|_\star^2 \leq |\varsigma| L_r c_\star^2 \|u_\varsigma\|_\star$. Using the upper bound on $\|u_\varsigma\|_\star$ for $\varsigma \to 0$ from Theorem 2.1 concludes the proof. □

If the solution u_0 of (2.20) is explicitly known, the following estimate becomes useful.

Proposition 2.5

u_ς, u_0 denote the solution of P_ς, P_0 respectively. Then, for $|\varsigma| < \frac{\lambda_{min}}{L_r c_\star^2}$, there holds

$$\left(\lambda_{min} - |\varsigma|\, L_r\, c_\star^2\right) \|u_\varsigma - u_0\|_\star \leq |\varsigma|\, L_r\, c_\star^2\, \|u_0\|_\star\,.$$

Proof

As in the proof of Proposition 2.4 we have
$\lambda_{min} \|u_\varsigma - u_0\|_\star^2 \leq |\varsigma| \int_\Omega |r(x, u_\varsigma)|\, |u_\varsigma - u_0|\, dx$. Using the triangle inequality for the right hand side yields

$$\lambda_{min} \|u_\varsigma - u_0\|_\star^2 \leq |\varsigma| \int_\Omega \left(|r(x, u_\rho) - r(x, u_0)| + |r(x, u_0)|\right) |u_\varsigma - u_0|\, dx\,.$$

Lipschitz-continuity of r and the Cauchy-Schwarz inequality give

$$\lambda_{min} \|u_\varsigma - u_0\|_\star^2 \leq |\varsigma|\, L_r\, \|u_\varsigma - u_0\|_{L^2(\Omega)}^2 + |\varsigma|\, L_r\, \|u_0\|_{L^2(\Omega)}\, \|u_\varsigma - u_0\|_{L^2(\Omega)}\,.$$

Using $\|\cdot\|_{L^2(\Omega)} \leq c_\star \|\cdot\|_\star$ yields the estimate. □

2.2.2. Error minimizing choice of the Poisson datum

To minimize the error in Proposition 2.5 for a fixed ς, we vary the Poisson datum in P_0 (2.20) and denote it by $f_\varsigma \in L^2(\Omega)$. In this case, the difference $u_\varsigma - u_0$ satisfies the equation $-\mathrm{div}(\Lambda\, \nabla(u_\varsigma - u_0)) = \varsigma\, r(\,\cdot\,, u_\varsigma) + f - f_\varsigma$. Using the same arguments as above we obtain $\|u_\varsigma - u_0\|_\star \leq \frac{c_\star}{\lambda_{min}} \|\varsigma\, r(\,\cdot\,, u_\varsigma) + f - f_\varsigma\|_{L^2(\Omega)}$. Now we set $f_\varsigma = f + \varsigma\, r(\,\cdot\,, \bar{u})$ for some constant $\bar{u} \in \mathbb{R}$. This and the Lipschitz continuity of r give $\lambda_{min} \|u_\varsigma - u_0\|_\star \leq c_\star\, |\varsigma|\, L_r\, \|u_\varsigma - \bar{u}\|_{L^2(\Omega)}$ and hence

$$\|u_\varsigma - u_0\|_\star \leq \frac{|\varsigma|\, c_\star\, L_r}{\lambda_{min} - |\varsigma|\, c_\star^2\, L_r} \|u_0 - \bar{u}\|_{L^2(\Omega)}\,. \tag{2.21}$$

The error minimizing $\bar{u}$ is given by the orthogonal projection of u_0 in $L^2(\Omega)$ on the subspace $\mathbb{R}$, i.e. by the mean value $\bar{u} := \frac{1}{|\Omega|} \int_\Omega u_0\, dx$ with $\|u_0 - \bar{u}\|_{L^2(\Omega)}^2 = \|u_0\|_{L^2(\Omega)}^2 - |\Omega|\, \bar{u}^2$.

On the other hand, $\bar{u}$ can be chosen suitably for a specific problem. E.g. we can set $\bar{u} = \frac{1}{|\partial\Omega|} \int_{\partial\Omega} u_0 \, d\sigma$ to treat (2.20) by boundary integral methods, see [25]. The error is controlled by (2.21) then.

2.3. Asymptotic behaviour in cylindrical domains

Now we treat the stationary problem (2.20) with a more specific geometry of Ω. We consider a cylinder $\Omega = \Omega_l := \Omega_{cr} \times (-l, l) \subset \mathbb{R}^d$ with a simply connected, open cross section $\Omega_{cr} \subset \mathbb{R}^{d-1}$ and a variable length $l > 0$. One expects that for large l the solution of (2.20) becomes independent of x_d - the axial coordinate of the cylinder in $x = (x_1, \ldots, x_d) \in \mathbb{R}^d$. Indeed it can be shown that under suitable assumptions the solution of (2.20) converges towards the extended solution of the associated cross-sectional problem in Ω_{cr}. To establish this convergence, we extend the method in [16] to boundary value problems with monotone boundary conditions. In this context we also refer to [27] who investigate halfspace asymptotics of semilinear elliptic equations.

2.3.1. Setting of the boundary value problems under investigation

The Neumann-boundary Γ_g decomposes in cross-sectional ends of the cylinder. I.e. $\Gamma_g = \Gamma_1 \cup \Gamma_2$ with $\Gamma_1 = \Omega_{cr} \times \{-l\}$, $\Gamma_2 = \Omega_{cr} \times \{l\}$. $\Gamma_\beta = \partial\Omega_{cr} \times [-l, l]$ denotes the monotone transmission boundary part.

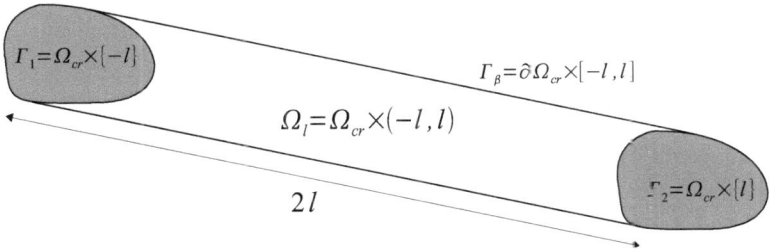

First we investigate the cross-sectional boundary value problem: Find $\bar{u} \in H^1(\Omega_{cr})$ such that

$$-\text{div}\left(\bar{\Lambda}\,\nabla \bar{u}\right) = \bar{f} \quad \text{in } \Omega_{cr} \qquad (2.22)$$
$$-\left(\bar{\Lambda}\,\nabla \bar{u}\right) n = \beta(\bar{u}) \quad \text{on } \partial\Omega_{cr}.$$

Assume $\bar{f} \in L^2(\Omega_{cr})$; $\bar{\Lambda} \in L^\infty(\Omega_{cr}, \mathbb{R}^{(d-1)\times(d-1)})$ with

$$\exists\,\bar{\lambda}_{min} > 0 : \inf_{x \in \Omega_{cr}} \left(y\,\Lambda(x)\,y\right) \geq \bar{\lambda}_{min} |y|^2,\ y \in \mathbb{R}^{d-1}.$$

$\beta : \mathbb{R} \to \mathbb{R}$ has the mapping properties described in (2.4). As in section 2.1.2, we define the physically consistent norm $\|v\|_{\star,\bar{cr}}^2 := \|\nabla v\|_{L^2(\Omega_{cr})}^2 + \frac{c_\beta}{\bar{\lambda}_{min}} \|v\|_{L^2(\partial\Omega_{cr})}^2$ which is equivalent to the canonical norm in $H^1(\Omega_{cr})$. Analogous to the proof of Theorem 2.1, (2.22) admits a unique solution.

Lemma 2.4
There exists a unique solution $\bar{u} \in H^1(\Omega_{cr})$ of (2.22) which is bounded by

$$\bar{\lambda}_{min} \|\bar{u}\|_{\star,\bar{cr}} \leq c_{\star,\bar{cr}} \|\bar{f}\|_{L^2(\Omega_{cr})} + \sqrt{\frac{|\partial\Omega_{cr}|\,\bar{\lambda}_{min}}{c_\beta}} |\beta(0)|,$$

where $c_{\star,\bar{cr}} = \sup_{v \in H^1(\Omega_{cr})\setminus\{0\}} \left(\|v\|_{L^2(\Omega_{cr})} / \|v\|_{\star,\bar{cr}}\right)$ *denotes the Friedrichs constant of Ω_{cr} w.r.t. $\bar{\lambda}_{min}$.*

Extension of the cross-sectional data to $\Omega_l \subset \mathbb{R}^d$

We extend the Poisson datum $\bar{f} \in L^2(\Omega_{cr})$ of (2.22) to $f_\infty \in L^2(\Omega_l)$ by

$$f_\infty(x_1, \ldots, x_{d-1}, x_d) = \bar{f}(x_1, \ldots, x_{d-1}) \quad \text{for } x_d \in (-l, l).$$

Analogously we extend the solution $\bar{u} \in H^1(\Omega_{cr})$ of (2.22) by

$$u_\infty(x_1, \ldots, x_{d-1}, x_d) = \bar{u}(x_1, \ldots, x_{d-1});$$

Hence $u_\infty \in H^1(\Omega_l)$. Finally we extend the conductivity/diffusivity matrix $\bar{\Lambda} \in L^\infty(\Omega_{cr}, \mathbb{R}^{(d-1)\times(d-1)})$ to $\Lambda_\infty \in L^\infty(\Omega_l, \mathbb{R}^{d\times d})$ via

$$\Lambda_\infty = \begin{pmatrix} \bar{\Lambda}(x_1,..,x_{d-1}) & \lambda_{\infty_{1,d}}(..,x_d) \\ & \vdots \\ 0 \ldots 0 & \lambda_{\infty_{d,d}}(..,x_d) \end{pmatrix} \qquad (2.23)$$

where $\lambda_{\infty_{k,d}} \in L^\infty(\Omega_l)$ denote the elements in the last column of Λ_∞. Observe that the construction in (2.23) fulfills the following consistency condition

$$(\Lambda_\infty \nabla u_\infty)(x_1, \ldots, x_d) = (\bar{\Lambda} \nabla \bar{u}, 0)(x_1, \ldots, x_{d-1}) \quad \text{for } x \in \Omega_l.$$

Thus the extension is isotropic in the axial direction. Note that this construction implies

$$\text{div}(\Lambda_\infty \nabla u_\infty)(x_1, \ldots, x_d) = \text{div}(\bar{\Lambda} \nabla \bar{u})(x_1, \ldots, x_{d-1}) \quad \text{for } x \in \Omega_l.$$

Suppose $\exists \lambda_{min} > 0 : \inf_{x \in \Omega_l}(y \Lambda(x) y) \geq \lambda_{min}|y|^2$, $y \in \mathbb{R}^d$. Observe that by the considerations above we have $0 < \lambda_{min} \leq \bar{\lambda}_{min}$. Therefore we introduce $\|v\|_{\star,cr}^2 := \|\nabla v\|_{L^2(\Omega_{cr})}^2 + \frac{c_\beta}{\lambda_{min}} \|v\|_{L^2(\partial\Omega_{cr})}^2$ and the associated Friedrichs constant becomes $c_{\star,cr} = \sup_{v \in H^1(\Omega_{cr})\setminus\{0\}} \left(\|v\|_{L^2(\Omega_{cr})} / \|v\|_{\star,cr} \right)$. On the other hand $\Lambda_\infty \in L^\infty(\Omega_l, \mathbb{R}^{d\times d})$ implies the existence of the upper bound $\lambda_{ddmax} := \text{ess sup}_{x \in \Omega_l} |\lambda_{\infty_{dd}}(x)|$. Consider now the cylinder boundary value problem: Find $u_l \in H^1(\Omega_l)$ such that

$$-\text{div}(\Lambda_\infty \nabla u_l) = f_\infty \quad \text{in } \Omega_l \qquad (2.24)$$
$$-(\Lambda_\infty \nabla u_l) n = \beta(u_l) \quad \text{on } \Gamma_\beta$$
$$(\Lambda_\infty \nabla u_l) n = g_1 \quad \text{on } \Gamma_1 \; ; \; (\Lambda_\infty \nabla u_l) n = g_2 \quad \text{on } \Gamma_2.$$

where $g_i \in H^{-1/2}(\Gamma_i)$, $i = 1, 2$ are given. The extended data have the same regularity properties as in (2.22). The equation is sub-resonant ($\varsigma = 0$). Hence, for every $l > 0$ there exists a unique solution $u_l \in H^1(\Omega_l)$ of (2.24) via

Theorem 2.1. By $\|\cdot\|_{\star,l}$ we denote the physically consistent norm on $H^1(\Omega_l)$, i.e. $\|v\|_{\star,l}^2 = \|\nabla v\|_{L^2(\Omega_l)}^2 + \frac{c_\beta}{\lambda_{min}} \|v\|_{L^2(\Gamma_\beta)}^2$.

2.3.2. Approximation of u_l by u_∞

In the following we want to compare the extended cross-sectional solution u_∞ with the cylindrical solution u_l for large l. To this end we complement M. Chipot's fundamental estimate in large cylinders [16].

Lemma 2.5 (tightening estimate)
Let u_l, u_∞ denote the solution of (2.24) and the extended solution of (2.22), respectively. Then, for $0 < l_2 < l_1 \leq l$ there holds

$$\|u_l - u_\infty\|_{\star,l_2} \leq \exp\left(\frac{-(l_1 - l_2)}{c_\lambda}\right) \|u_l - u_\infty\|_{\star,l_1} \; ; \; c_\lambda = \frac{C_{\star,cr} \lambda_{ddmax}}{\lambda_{min}}. \quad (2.25)$$

Proof
The difference $(u_l - u_\infty) \in H^1(\Omega_l)$ solves

$$\begin{aligned}
-\operatorname{div}(\Lambda_\infty \nabla (u_l - u_\infty)) &= 0 & &\text{in } \Omega_l \\
-\Lambda_\infty \nabla (u_l - u_\infty) n &= \beta(u_l) - \beta(u_\infty) & &\text{on } \Gamma_\beta \\
\Lambda_\infty \nabla (u_l - u_\infty) n &= g_1 & &\text{on } \Gamma_1 \; ; \; \Lambda_\infty \nabla (u_l - u_\infty) n = g_2 \quad \text{on } \Gamma_2
\end{aligned}$$

and thus the associated weak form for all $v \in H^1(\Omega_l)$ reads as

$$\int_{\Omega_l} \Lambda_\infty \nabla(u_l - u_\infty) \nabla v \, dx + \int_{\Gamma_\beta} (\beta(u_l) - \beta(u_\infty)) v \, d\sigma \quad (2.26)$$
$$- \int_{\Gamma_1} g_1 v \, d\sigma - \int_{\Gamma_2} g_2 v \, d\sigma = 0.$$

We introduce a piecewise linear truncating function $\gamma : \Omega_l \to [0,1]$ which is constant w.r.t. $(x_1, \ldots, x_{d-1})$ and

$$\gamma(x_d) = \begin{cases} 1 & \text{in } (-l_2, l_2) \\ 0 & \text{in } (-l, l) \setminus [-l_1, l_1] \end{cases} \; ; \; \text{i.e.} \; \left|\frac{\partial \gamma}{\partial x_d}\right| \leq \frac{1}{l_1 - l_2}.$$

Setting $v = (u_l - u_\infty)\gamma$ we obtain for the first summand in (2.26)

$$\int_{\Omega_l} \Lambda_\infty \nabla(u_l - u_\infty) \nabla v \, dx = \int_{\Omega_{l_1}} \Lambda_\infty \nabla(u_l - u_\infty) \nabla((u_l - u_\infty)\gamma) \, dx$$

$$\geq \lambda_{min} \int_{\Omega_{l_2}} |\nabla(u_l - u_\infty)|^2 \, dx$$

$$+ \int_{\Omega_{l_1} \setminus \Omega_{l_2}} \Lambda_\infty \nabla(u_l - u_\infty) \begin{pmatrix} 0 \\ \vdots \\ 0 \\ \partial_{x_d}\gamma \end{pmatrix} (u_l - u_\infty) \, dx.$$

v vanishes on Γ_1 and Γ_2; we use the monotonicity of β and the definition of γ to estimate the remaining part of (2.26)

$$\int_{\Gamma_\beta} (\beta(u_l) - \beta(u_\infty)) v \, d\sigma = \langle \beta(u_l) - \beta(u_\infty), (u_l - u_\infty)\gamma \rangle_{L^2(\Gamma_\beta \cap \Omega_{l_1})}$$

$$\geq c_\beta \|u_l - u_\infty\|^2_{L^2(\Gamma_\beta \cap \Omega_{l_2})}.$$

Thus we have

$$\lambda_{min} \int_{\Omega_{l_2}} |\nabla(u_l - u_\infty)|^2 \, dx + c_\beta \|u_l - u_\infty\|^2_{L^2(\Gamma_\beta \cap \Omega_{l_2})}$$

$$\leq - \int_{\Omega_{l_1} \setminus \Omega_{l_2}} \Lambda_\infty \nabla(u_l - u_\infty) \begin{pmatrix} C \\ \vdots \\ C \\ \partial_{x_c}\gamma \end{pmatrix} (u_l - u_\infty) \, dx.$$

The definitions of Λ_∞, $\|\cdot\|_{\star, l_2}$ and γ imply

$$\|u_l - u_\infty\|^2_{\star, l_2} \leq \frac{\lambda_{min}^{-1}}{l_1 - l_2} \int_{\Omega_{l_1} \setminus \Omega_{l_2}} |\lambda_{\infty d,d} \partial_{x_d}(u_l - u_\infty)| \, |u_l - u_\infty| \, dx$$

$$\leq \frac{\lambda_{ddmax} \lambda_{min}^{-1}}{l_1 - l_2} \int_{\Omega_{l_1} \setminus \Omega_{l_2}} |\partial_{x_d}(u_l - u_\infty)| \, |u_l - u_\infty| \, dx$$

$$\leq \frac{\lambda_{ddmax}\lambda_{min}^{-1}}{l_1-l_2} \int_{\Omega_{l_1}\setminus\Omega_{l_2}} \left(\frac{c_{\star,cr}}{2}|\partial_{x_d}(u_l-u_\infty)|^2 + \frac{1}{2c_{\star,cr}}|u_l-u_\infty|^2\right)dx, \quad (2.27)$$

where the last estimate follows from Young's inequality. By the definition of Friedrichs constant $c_{\star,cr}$ there holds for a.e. x_d

$$\int_{\Omega_{cr}} (u_l-u_\infty)^2\, dx_1\cdots dx_{d-1} \leq$$

$$c_{\star,cr}^2 \left(\int_{\Omega_{cr}} \sum_{i=1}^{d-1}(\partial_{x_i}(u_l-u_\infty))^2\, dx_1\cdots dx_{d-1} + \frac{c_\beta}{\lambda_{min}} \int_{\partial\Omega_{cr}} ((u_l-u_\infty))^2\, ds\right)$$

and an integration over $x_d \in (-l_1,l_1)\setminus(-l_2,l_2)$ yields

$$\int_{\Omega_{l_1}\setminus\Omega_{l_2}} (u_l-u_\infty)^2\, dx \leq$$

$$c_{\star,cr}^2 \left(\int_{\Omega_{l_1}\setminus\Omega_{l_2}} \sum_{i=1}^{d-1}(\partial_{x_i}(u_l-u_\infty))^2\, dx + \frac{c_\beta}{\lambda_{min}} \int_{\Gamma_\beta\cap(\Omega_{l_1}\setminus\Omega_{l_2})} (u_l-u_\infty)^2\, d\sigma\right).$$

Inserting this inequality in (2.27) gives

$$\|u_l-u_\infty\|_{\star,l_2}^2 \leq \frac{c_\lambda}{2(l_1-l_2)} \int_{\Omega_{l_1}\setminus\Omega_{l_2}} (\nabla(u_l-u_\infty))^2\, dx$$

$$+ \frac{c_\lambda \lambda_{min}^{-1} c_\beta}{2(l_1-l_2)} \int_{\Gamma_\beta\cap(\Omega_{l_1}\setminus\Omega_{l_2})} (u_l-u_\infty)^2\, d\sigma$$

$$=: \frac{c_\lambda}{2(l_1-l_2)} \|u_l-u_\infty\|_{\star,l_1\setminus l_2}^2 \; ; \; c_\lambda = \frac{c_{\star,cr}\lambda_{ddmax}}{\lambda_{min}} \quad (2.28)$$

For a fixed $l > 0$ we define the mapping $\mathcal{F}: (0,l) \to \mathbb{R}$ with $\mathcal{F}(s) := \|u_l-u_\infty\|_{\star,s}^2$. $\mathcal{F}$ is a.e. differentiable and by (2.28) we have

$$\mathcal{F}(l_2) \leq \frac{c_\lambda}{2}\frac{\mathcal{F}(l_1)-\mathcal{F}(l_2)}{l_1-l_2} \xrightarrow[l_1\to l_2]{} \frac{c_\lambda}{2}\mathcal{F}'(l_2).$$

Multiplying this relation by $\exp\left(-2\,c_\lambda^{-1}\,s\right)$ and using the product rule, we get $\left(\exp\left(-2\,c_\lambda^{-1}\,s\right)\mathcal{F}(s)\right)' \geq 0$ i.e. the mapping $s \mapsto \exp\left(-2\,c_\lambda^{-1}\,s\right)\mathcal{F}(s)$ is monotonically increasing. An evaluation of the monotonicity for $l_2 < l_1$ implies the assertion. $\square$

Theorem 2.3
With the notation of Lemma 2.5 and $l > l_2$ there holds

$$\lambda_{min}\,\|u_l - u_\infty\|_{*,l_2} \leq \exp\left(\frac{-(l-l_2)}{c_\lambda}\right)\left(C_1\,\|g_1\|_{L^2(\Gamma_1)} + C_2\,\|g_2\|_{L^2(\Gamma_2)}\right)$$

where $C_i := \|\tau\|_{tr} = \sup_{\|v\|_{*,l_2}\leq 1}\|\tau(v)\|_{L^2(\Gamma_i)}$ denotes the norm of the trace map $\tau : H^1(\Omega_{l_2}) \to L^2(\Gamma_i)$.

Proof
$(u_l - u_\infty) \in H^1(\Omega_l)$ satisfies (2.26). Set $v = u_l - u_\infty$ and the monotonicity condition for β gives $\|u_l - u_\infty\|_{*,l}^2 \leq \frac{1}{\lambda_{min}}\left(\int_{\Gamma_1} g_1\,(u_l - u_\infty)\,d\sigma + \int_{\Gamma_2} g_2\,(u_l - u_\infty)\,d\sigma\right)$ which implies $\|u_l - u_\infty\|_{*,l} \leq \frac{1}{\lambda_{min}}\left(C_1\,\|g_1\|_{L^2(\Gamma_1)} + C_2\,\|g_2\|_{L^2(\Gamma_2)}\right)$. Using Lemma 2.5 with $l_1 = l$ concludes the proof. $\square$

2.3.3. Locality of the convergence

We show by a counterexample, that Theorem 2.3 cannot be extended to global convergence, i.e. $\|u_l - u_\infty\|_{*,l} \not\xrightarrow[l\to\infty]{} 0$, in general.

The cross-sectional problem

For $d = 2$, $\bar{\Lambda} = 1$, $\beta(s) = s$ (i.e. $\bar{\lambda} = c_\beta = 1$), and $\bar{f} = 1$ we treat the cross-sectional problem

$$-\bar{u}'' = 1 \quad \text{in } (0,1) := \Omega_{cr};\quad \bar{u}'(0) = \bar{u}(0),\quad -\bar{u}'(1) = \bar{u}(1). \quad (2.29)$$

The unique solution $\bar{u} \in H^1(\Omega_{cr})$ is given by $\bar{u}(x_1) = \frac{1}{2}(-x_1^2 + x_1 + 1)$.

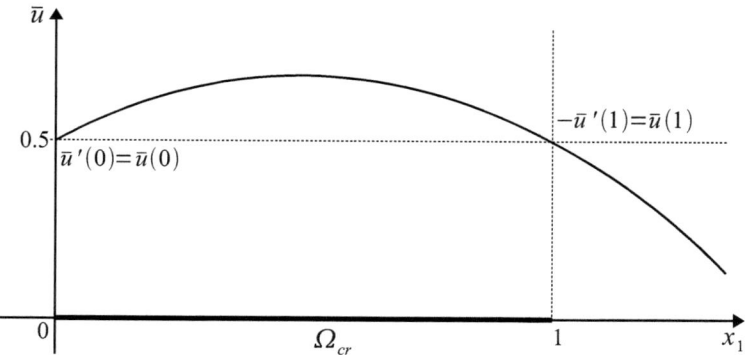

The cylinder problem

Consider now the associated boundary value problem in $\Omega_l = (0,1) \times (-l, l)$ where the data is extended by means of section 2.3.1. We have
$\Lambda_\infty = \begin{pmatrix} 1 & 0 \\ 0 & 1 \end{pmatrix}$, $f_\infty = 1$ and $g_1 = 1$ on $\Gamma_1 = \{(x_1, x_2) \in \mathbb{R}^2, x_2 = -l\}$; $g_2 = 1$ on $\Gamma_2 = \{(x_1, x_2) \in \mathbb{R}^2, x_2 = l\}$ and thus

$$-\Delta u_l = 1 \quad \text{in} \quad \Omega_l$$
$$\frac{\partial u_l}{\partial x_1}(0, x_2) = u_l(0, x_2) \ , \ -\frac{\partial u_l}{\partial x_1}(1, x_2) = u_l(1, x_2) \ ; \ x_2 \in (-l, l) \quad (2.30)$$
$$-\frac{\partial u_l}{\partial x_2}(x_1, -l) = 1 \ , \ \frac{\partial u_l}{\partial x_2}(x_1, l) = 1 \ ; \ x_1 \in (0, 1).$$

To solve this problem we make the ansatz $u_l = \bar{u}\, w$, where $\bar{u}$ denotes the solution of the cross-sectional problem (2.29). This decomposition and the monotone boundary condition for $\bar{u}$ imply the associated condition in (2.30). w has to solve the following remaining ordinary differential equation

$$\ddot{w} = \frac{1}{\bar{u}}(w - 1) \text{ in } (-l, l); \ \dot{w}(-l) = -\frac{1}{\bar{u}}, \ \dot{w}(l) = \frac{1}{\bar{u}} \quad (2.31)$$

where " $\cdot$ " denotes the derivative w.r.t. x_2. Observe that by previous investigations $\bar{u}$ is positive in $\Omega_{cr} = (0, 1)$ and constant w.r.t. x_2 such that (2.31) is well defined. The solution reads as $w(x_2) = 1 + \sqrt{\frac{1}{\bar{u}}} \frac{\cosh(\sqrt{\frac{1}{\bar{u}}} x_2)}{\sinh(\sqrt{\frac{1}{\bar{u}}} l)}$ and we obtain

$$u_l(x_1, x_2) = \bar{u}(x_1) \left(1 + \sqrt{\frac{1}{\bar{u}(x_1)}} \frac{\cosh\left(\sqrt{\frac{1}{\bar{u}(x_1)}} x_2\right)}{\sinh\left(\sqrt{\frac{1}{\bar{u}(x_1)}} l\right)} \right).$$

We depict u_l for $l = 8$ in the following figure.

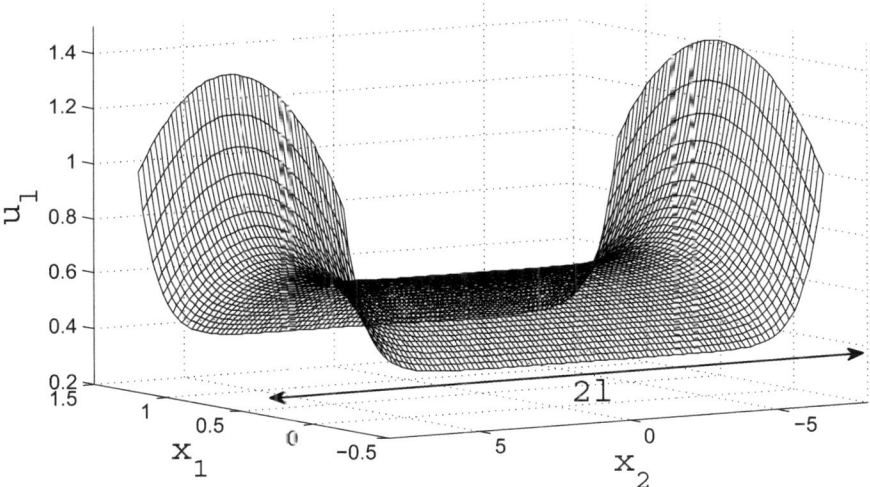

For a fixed $l_0 > 0$ we verify the local convergence $\|u_l - u_\infty\|_{z,l_0} \xrightarrow[l \to \infty]{} 0$ asserted by Theorem 2.3. There holds

$$\nabla(u_l - u_\infty) = \begin{pmatrix} \frac{\omega^2 \bar{u}'}{2 \sinh(\omega l)} \left(\left(\frac{1}{\omega} + l\right) \tanh(\omega l)^{-1} \cosh(\omega x_2) - x_2 \sinh(\omega x_2) \right) \\ \frac{\sinh(\omega x_2)}{\sinh(\omega l)} \end{pmatrix}$$

where $\omega = \sqrt{\frac{1}{\bar{u}}}$. Due to the properties of $\bar{u}$ we have $\sqrt{\frac{\xi}{\alpha}} \le \omega \le \sqrt{2}$ and $|\bar{u}'| \le \frac{1}{2}$. This implies for $l > 1$

$$|\nabla(u_l - u_\infty)|^2 \le \frac{1}{\sinh(\omega l)^2} \left(\left(\frac{1}{\omega} + \frac{|x_2|}{2} + l + 1 \right) \cosh(\omega x_2) \right)^2.$$

The traces on Γ_β read as

$$(u_l - u_\infty)(0, x_2) = (u_l - u_\infty)(1, x_2) = \sqrt{\frac{1}{2} \frac{\cosh(\sqrt{2}\, x_2)}{\sinh(\sqrt{2}\, l)}}, \quad x_2 \in (-l_0, l_0).$$

A rough estimate for $l > l_0$ yields

$$\|\nabla (u_l - u_\infty)\|_{L^2(\Omega_l)} \leq 8\sqrt{2\, l_0}\, (1+l)\, \exp(2\, l_0 - l)$$
$$\|u_l - u_\infty\|_{L^2(\Gamma_\beta)} \leq 8\sqrt{2\, l_0}\, \exp\left(\sqrt{2}\,(l_0 - l)\right)$$

and hence $\|u_l - u_\infty\|_{\star,l_0} \leq 16\sqrt{l_0}\, (1+l)\, \exp(2\, l_0 - l)$.

Lower bound for $\|u_l - u_\infty\|_{\star,l}$

It suffices to find a lower bound for the boundary part of $\|\cdot\|_{\star,l}$. We have

$$\|u_l - u_\infty\|_{\star,l}^2 \geq \|u_l - u_\infty\|_{L^2(\Gamma_\beta)}^2 = \frac{\int_{-l}^{l} \left(e^{\sqrt{2}\, x_2} + e^{-\sqrt{2}\, x_2}\right)^2 \, dx_2}{4\sinh(\sqrt{2}\, l)^2}$$
$$\geq \sqrt{\frac{1}{2} \frac{\sinh(2\sqrt{2}\, l)}{4\sinh(\sqrt{2}\, l)^2}} \geq \sqrt{\frac{1}{8}}$$

which implies $\|u_l - u_\infty\|_{\star,l} \not\to 0$ as $l \to \infty$.

2.4. An estimate for $c_\star$ in star-shaped domains

To identify the sub-resonance condition in Theorem 2.1 and the estimates in Proposition 2.5 and 2.3 by geometrical and physical parameters, we need an estimate for $c_\star$.

2.4.1. Preliminary remarks

In order to relate the estimate with classical results, we give a brief overview about optimal constants in Friedrichs and Poincaré inequalities. They are associated with the Dirichlet and the Neumann eigenvalue problem for the Laplacian, respectively.

Dirichlet eigenvalues and Friedrichs constant $c_F(\Omega)$

Consider the Dirichlet problem

$$-\Delta u = \lambda u \text{ in } \Omega \, ; \, u = 0 \text{ on } \Gamma. \tag{2.32}$$

This eigenvalue problem has a discrete spectrum of Dirichlet eigenvalues $(\lambda_k)_{k\in\mathbb{N}} \subset \mathbb{R}$ with $0 < \lambda_1 \leq \ldots \leq \lambda_k \xrightarrow{k\to\infty} \infty$. The Pólya Conjecture [67] identifies the Weyl asymptotics [77] $\lambda_k \sim \frac{4\pi^2 k^{2/d}}{(\omega_d |\Omega|)^{2/d}}$ as the lower bound $\lambda_k \geq \frac{4\pi^2 k^{2/d}}{(\omega_d |\Omega|)^{2/d}}$, where $\omega_d = \frac{\pi^{d/2}}{\Gamma(d/2+1)}$ denotes the volume of the unit ball in $\mathbb{R}^d$. Up to date this conjecture is not proved for arbitrary bounded Lipschitz domains Ω. Therefore we use the -to date- best proven result [52] with $\sum_{j=1}^k \lambda_j \geq \frac{d}{d+2} \frac{4\pi^2 k^{(d+2)/d}}{(\omega_d |\Omega|)^{2/d}}$ which includes $\lambda_1 \geq \frac{d}{d+2} \frac{4\pi^2}{(\omega_d |\Omega|)^{2/d}}$. Using the variational formulation of (2.32) we obtain $\lambda_1 \leq \frac{\|\nabla u\|_{L^2(\Omega)}^2}{\|u\|_{L^2(\Omega)}^2}$, $u \in H_0^1(\Omega) \setminus \{0\}$. Thus the principal Dirichlet eigenvalue λ_1 provides an optimal constant $c_F(\Omega) = \frac{1}{\sqrt{\lambda_1}}$ in Friedrichs inequality

$$\|u\|_{L^2(\Omega)} \leq c_F(\Omega) \|\nabla u\|_{L^2(\Omega)} \, , \, u \in H_0^1(\Omega)$$

via $c_F(\Omega) = \sup_{u \in H_0^1(\Omega)\setminus\{0\}} \frac{\|u\|_{L^2(\Omega)}}{\|\nabla u\|_{L^2(\Omega)}}$ and the upper bound $c_F(\Omega) \leq \sqrt{\frac{d+2}{d}} \frac{\sqrt[d]{\omega_d |\Omega|}}{2\pi}$.

Neumann eigenvalues and Poincaré constant $c_P(\Omega)$

Consider now the Neumann eigenvalue problem for the Laplacian:

$$-\Delta u = \mu u \text{ in } \Omega \, ; \, \frac{\partial u}{\partial n} = 0 \text{ on } \Gamma. \tag{2.33}$$

We obtain a discrete spectrum $(\mu_k)_{k\in\mathbb{N}}$ with $0 = \mu_1 \leq \mu_2 \leq \ldots \leq \mu_k \xrightarrow{k\to\infty} \infty$. The eigenfunctions for $\mu_1 = 0$ are constant. (2.33) and Gauß' Divergence

Theorem imply $\int_\Omega \mu_k u \, dx = 0$. I.e. the requirement $\frac{1}{|\Omega|} \int_\Omega u \, dx = 0$ for eigenfunctions yields $\mu_k > 0$, $k \geq 2$. Using this in the variational formulation of (2.33) we obtain the first nonvanishing eigenvalue μ_2 with $\mu_2 =$ inf $\left\{ \frac{\|\nabla u\|^2_{L^2(\Omega)}}{\|u\|^2_{L^2(\Omega)}} \,,\, u \in H^1_{m_0}(\Omega) \setminus \{0\} \right\}$
where $H^1_{m_0}(\Omega) := \left\{ u \in H^1(\Omega) \,:\, \frac{1}{|\Omega|} \int_\Omega u \, dx = 0 \right\}$. An optimal lower bound for μ_2 in convex domains is given by $\mu_2 \geq \frac{\pi^2}{diam(\Omega)^2}$ [66]. Thus μ_2 provides an optimal constant in the Poincaré inequality

$$\|u\|_{L^2(\Omega)} \leq c_P(\Omega) \, \|\nabla u\|_{L^2(\Omega)} \,,\, u \in H^1_{m_0}(\Omega)$$

via $c_P(\Omega) = \frac{1}{\sqrt{\mu_2}}$ and the optimal bound for convex domains $c_P(\Omega) \leq \frac{diam(\Omega)}{\pi}$.

Dirichlet-Neumann comparison, Isodiametric inequality

Based on considerations in [33], N. Filonov [30] proved the estimate between Dirichlet and Neumann eigenvalues $\mu_{k+1} < \lambda_k$ i.e. $c_F(\Omega) \leq c_P(\Omega)$ for arbitrary Lipschitz Domains. The direct result in [52] combined with the isodiametric inequality gives $c_F(\Omega) < \sqrt{\frac{d+2}{d}} \frac{\omega_d^{2/d}}{4\pi} \, diam(\Omega)$. The Dirichlet-Neumann comparison and [66] gives for convex domains $c_F(\Omega) \leq \frac{diam(\Omega)}{\pi}$. In convex domains, the second -indirect- estimate is slightly better for $d = 2$. In dimension $d \geq 3$ and for arbitrary Lipschitz domains the direct estimate should be preferred.

2.4.2. Inhomogeneous Friedrichs inequality in $W^{1,p}(\Omega)$

Such as in section 2.1.1 we assume that the boundary Γ decomposes in Γ_β and Γ_g with $\Gamma_g \cap \Gamma_\beta = \emptyset$ and $\overline{\Gamma}_g \cup \overline{\Gamma}_\beta = \Gamma$.

Theorem 2.4
Assume that $\Omega \subset \mathbb{R}^d$ is star-shaped such that every $x_0 \in \Gamma_\beta$ is a center of Ω and $p \in [1, \infty)$. Then, for $u \in W^{1,p}(\Omega)$ we have

$$\|u\|^p_{L^p(\Omega)} \leq 2^{p-1} \left(\frac{diam(\Omega)}{d} \frac{|\Gamma|}{|\Gamma_\beta|} \|u\|^p_{L^p(\Gamma_\beta)} + \frac{diam(\Omega)^p}{p} \|\nabla u\|^p_{L^p(\Omega)} \right).$$

Proof
Choose $x_0 \in \Gamma_\beta$ such that

$$u(x_0)^p \leq \left(\frac{1}{|\Gamma_\beta|} \int_{\Gamma_\beta} u \, d\sigma\right)^p \leq \frac{1}{|\Gamma_\beta|} \int_{\Gamma_\beta} u^p \, d\sigma = \frac{1}{|\Gamma_\beta|} \|u\|_{L^p(\Gamma_\beta)}^p \qquad (2.34)$$

where the second estimate follows by Jensen's inequality, see (A.5).

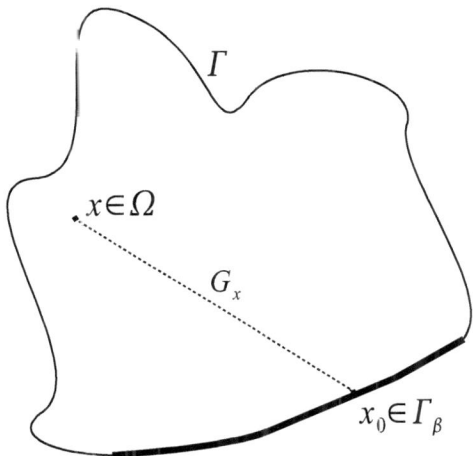

Assume $u \in C^1(\Omega) \cap C(\overline{\Omega})$ and $0 = x_0 \in \Gamma_\beta$. Otherwise use the translation $\tilde{x} := x - x_0$. For every $x \in \Omega$ we define the line segment $L_x = \{t\,x\,;\, t \in (0,1)\} \subset \Omega$. Then there holds

$$\begin{aligned} u(x) - u(0) &= \int_0^1 \frac{d(u \circ \gamma)(t)}{dt} \, dt = \int_0^1 \langle \nabla u(\gamma(t)) \cdot \dot{\gamma}(t) \rangle \, dt \\ &\leq \int_0^1 |\nabla u(\gamma(t))| \, |\dot{\gamma}(t)| \, dt = \int_{L_x} |\nabla u| \, d\gamma \end{aligned}$$

where $\gamma : [0,1] \to L_x$ denotes a parametrization of L_x.
I.e. $u(x) \leq u(0) + \int_{L_x} |\nabla u| \, d\gamma$. Jensen's inequality applied twice gives

$$u(x)^p \leq 2^{p-1}\left(u(0)^p + |L_x|^{p-1} \int_{L_x} |\nabla u|^p \, d\gamma\right). \qquad (2.35)$$

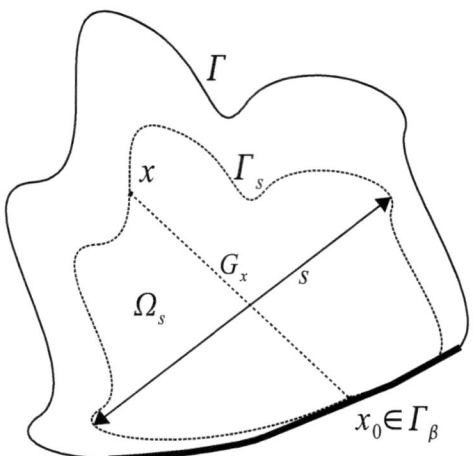

By $\Omega_s := \frac{s}{diam(\Omega)} \Omega$, $s \in (0, diam(\Omega))$, with its boundary Γ_s, we denote a homotopic contraction of Ω to $x_0 = 0$. This contraction exists since Ω is star shaped and thus contractible, [62]. Since $s = diam(\Omega_s)$, we have $|\Gamma_s| = \frac{|\Gamma|}{diam(\Omega)^{d-1}} s^{d-1}$.

As $x \in \Gamma_s$ implies $|L_x| = |x| \leq s$ and since Ω_s is star shaped with center $0 \in \Gamma_\beta$, an integration of (2.35) over Γ_s yields

$$\int_{\Gamma_s} u^p \, d\sigma \leq 2^{p-1} \left(|\Gamma_s| u(0)^p + s^{p-1} \int_{\Omega_s} |\nabla u|^p \, dx \right).$$

Using $\int_{\Omega_s} |\nabla u|^p \, dx \leq \|\nabla u\|_{L^p(\Omega)}^p$, an integration over s provides via Cavalieri's principle (A.3)

$$\|u\|_{L^p(\Omega)}^p \leq 2^{p-1} \left(\frac{|\Gamma|}{diam(\Omega)^{d-1}} \int_0^{diam(\Omega)} s^{d-1} \, ds \, |u(0)|^p + \frac{diam(\Omega)^p}{p} \|\nabla u\|_{L^p(\Omega)}^p \right).$$

Now (2.34) and an extension via density to arbitrary $u \in W^{1,p}(\Omega)$ finally imply the assertion. □

In the following we want to extract a uniform constant $c_\star$ for the inequality $\|v\|_{L^2(\Omega)} \leq c_\star \|v\|_\star$ via Theorem 2.4. This and the definition of $\|\cdot\|_\star$ give rise to distinguish between a small scale case $diam(\Omega) \leq \frac{2\,|\Gamma|\lambda_{min}}{d\,|\Gamma_\beta|\,c_\beta}$ and a large scale case $diam(\Omega) > \frac{2\,|\Gamma|\lambda_{min}}{d\,|\Gamma_\beta|\,c_\beta}$.

Proposition 2.6 (An estimate for $c_\star$ via scaling)
Under the conditions of Theorem 2.4 and $p = 2$. we have for every $u \in H^1(\Omega)$ the small scale case $diam(\Omega) \leq \frac{2\,|\Gamma|\lambda_{min}}{d\,|\Gamma_\beta|\,c_\beta}$ which implies

$$\|u\|^2_{L^2(\Omega)} \leq \frac{2}{d}\,\frac{|\Gamma|}{|\Gamma_\beta|}\,\frac{\lambda_{min}}{c_\beta}\,diam(\Omega)\,\|u\|^2_\star \,;\quad \text{I.e. } c_\star \leq \sqrt{\frac{2}{d}\,\frac{|\Gamma|}{|\Gamma_\beta|}\,\frac{\lambda_{min}}{c_\beta}\,diam(\Omega)}\,.$$

or the large scale case $diam(\Omega) > \frac{2\,|\Gamma|\lambda_{min}}{d\,|\Gamma_\beta|\,c_\beta}$ which implies

$$\|u\|_{L^2(\Omega)} \leq diam(\Omega)\,\|u\|_\star \,;\quad \text{I.e. } c_\star \leq diam(\Omega)\,.$$

These bounds for $c_\star$ will be identified with physical quantities in chapter 2.

Remark
Note that the case distinction in Proposition 2.6 yields estimates for $c_\star$ which are not optimal. In order to obtain an orientation for the accuratesse of the estimate we compare it with the more precise estimates for

Subresonant states of homogeneous dirichlet problems.

We consider the special case $\Gamma = \Gamma_\beta, d = 2$ and investigate the subresonant state of the homogeneous Dirichlet-problem

$$\begin{aligned} -\operatorname{div}(\Lambda\nabla u) &= \varsigma\,r(\,\cdot\,,u) + f \quad \text{in } \Omega \\ u &= 0 \quad \text{in } \partial\Omega \end{aligned} \qquad (2.36)$$

where Λ, r, , f fulfill the same condititons as in (2.5).

Proposition 2.7
Let Ω be a convex domain and $|\varsigma| < \frac{\lambda_{min}\pi^2}{L_r\,diam(\Omega)^2}$. Then, for all $f \in L^2(\Omega)$ there

exists a unique solution $u \in H_0^1(\Omega)$ of (2.36) which is bounded by

$$\left(\lambda_{min} - L_r |\varsigma| \left(\frac{diam(\Omega)}{\pi}\right)^2\right) \|u\|_{H_0^1(\Omega)} \leq C_f + C_r$$

where $\|u\|_{H_0^1(\Omega)} := \|\nabla u\|_{L^2(\Omega)}$
and $C_f = \frac{diam(\Omega)}{\pi} \|f\|_{L^2(\Omega)}$, $C_r = |\varsigma| \frac{diam(\Omega)}{\pi} \|r(\,\cdot\,, 0)\|_{L^2(\Omega)}$.

Proof
The variational form of (2.36) reads as $\langle Au, v \rangle = \langle b, v \rangle \; \forall v \in H_0^1(\Omega)$

$$\langle Au, v \rangle := \int_\Omega \nabla u \wedge \nabla v \, dx - \int_\Omega \varsigma \, r(x, u) \, v \, dx$$
$$\langle b, v \rangle := \int_\Omega f v \, dx.$$

We show the strong monotonicity of the operator $A : H_0^1(\Omega) \to H^{-1}(\Omega)$.

$$\begin{aligned}
\langle Au - Av, u - v \rangle &\geq \lambda_{min} \|u - v\|_{H_0^1(\Omega)}^2 - |\varsigma| L_r \|u - v\|_{L^2(\Omega)}^2 \\
&\geq (\lambda_{min} - |\varsigma| L_r c_F^2) \|u - v\|_{H_0^1(\Omega)}^2
\end{aligned}$$

Here we identify the Friedrichs constant for convex domains in $\mathbb{R}^2$ via the considerations in section 2.4.1 by $c_F = \frac{diam(\Omega)}{\pi}$.

The hemicontinuity of A as well as the boundedness of the linear form $b \in H^{-1}(\Omega)$ are clear. Thus existence and uniqueness follow by the Theorem of Browder and Minty. For the bound on the solution we note that

$$\begin{aligned}
\langle Au, u \rangle &\geq \lambda_{min} \|u\|_{H_0^1(\Omega)}^2 - |\varsigma| \langle r(\,\cdot\,, u), u \rangle \\
&\geq (\lambda_{min} - |\varsigma| L_r c_F^2) \|u\|_{H_0^1(\Omega)}^2 - |\varsigma| |\langle r(\,\cdot\,, 0), u \rangle| \\
&\geq (\lambda_{min} - |\varsigma| L_r c_F^2) \|u\|_{H_0^1(\Omega)}^2 - |\varsigma| c_F \|r(\,\cdot\,, 0)\|_{L^2(\Omega)} \|u\|_{H_0^1(\Omega)}.
\end{aligned}$$

On the other hand we have

$$\langle b, u \rangle \leq c_F \|f\|_{L^2(\Omega)} \|u\|_{H_0^1(\Omega)}$$

which yields the assertion. □

Comparison with the subresonant state of (2.5)
We assume $p = d = 2$. In the appropriate large scale case $diam(\Omega) \geq \frac{\lambda_{min}}{c_\beta}$ the s.r.s. of (2.5) is given by $|\varsigma| \leq \frac{\lambda_{min}}{L_r \, diam(\Omega)^2}$.
We see from Proposition 2.7 that this is fairly related to the subresonant state in the homogeneous Dirichlet problem (2.36) $|\varsigma| \leq \frac{\lambda_{min} \pi^2}{L_r \, diam(\Omega)^2}$ where we assume more restrictively the convexity of the domain Ω and $u|_\Gamma = 0$.

2.4.3. Estimate for the trace embedding $W^{1,p}(\Omega) \hookrightarrow L^p(\Gamma_g)$

We use the method of proof of Theorem 2.4 to get an estimate for the embedding between $W^{1,p}(\Omega)$ and $L^p(\Gamma_g)$. In particular, for $p = 2$ we are able to identify the trace embedding constant c_{L^2} from Theorem 2.1.

Corollary 2.5
Assume that $\Omega \subset \mathbb{R}^d$ is star-shaped such that every $x_0 \in \Gamma_\beta$ is a center of Ω and $p \in [1, \infty)$. Then, for $u \in W^{1,p}(\Omega)$ we have

$$\|u\|^p_{L^p(\Gamma_g)} \leq 2^{p-1} \left(\frac{|\Gamma_g|}{|\Gamma_\beta|} \|u\|^p_{L^p(\Gamma_\beta)} + diam(\Omega)^{p-1} \|\nabla u\|^p_{L^p(\Omega)} \right).$$

Proof
We follow the arguments in the Proof of Theorem 2.4 till relation (2.35). Now we integrate over Γ_g and obtain

$$\int_{\Gamma_g} u^p \, d\sigma \leq 2^{p-1} \left(|\Gamma_g| \, u(0)^p + diam(\Omega)^{p-1} \int_\Omega |\nabla u|^p \, dx \right).$$

Thus (2.34) and an extension via density to arbitrary $u \in W^{1,p}(\Omega)$ yields the assertion. □

Using a distinction between a small scale and a large scale case we obtain

Proposition 2.8 (An estimate for c_{L^2} via scaling)
Under the conditions of Corollary 2.5 and $p = 2$ we have for every $u \in H^1(\Omega)$ the small scale case $diam(\Omega) \leq \frac{|\Gamma_g| \lambda_{min}}{|\Gamma_\beta| c_\beta}$ which implies

$$\|u\|_{L^2(\Gamma_g)}^2 \leq \frac{2|\Gamma_g|}{|\Gamma_\beta|} \frac{\lambda_{min}}{c_\beta} \|u\|_\star^2 \ ; \quad I.e. \ c_{L^2} \leq \sqrt{\frac{2|\Gamma_g|}{|\Gamma_\beta|} \frac{\lambda_{min}}{c_\beta}} \ .$$

or the large scale case $diam(\Omega) > \frac{|\Gamma_g|\lambda_{min}}{|\Gamma_\beta|c_\beta}$ which implies

$$\|u\|_{L^2(\Gamma_g)} \leq \sqrt{2\,diam(\Omega)} \ \|u\|_\star \ ; \quad I.e. \ c_{L^2} \leq \sqrt{2\,diam(\Omega)} \ .$$

Embedding inequality w.r.t. the canonical norm in $W^{1,p}(\Omega)$

The method in the proof of Theorem 2.4 yields also an estimate w.r.t. the canonical norm in $W^{1,p}(\Omega)$. We inforce the assumptions on Ω, since we need that any line segment between two points in Ω has to be included in Ω.

Corollary 2.6
Assume that $\Omega \subset \mathbb{R}^d$ is convex and $p \in [1,\infty)$. Then, for $u \in W^{1,p}(\Omega)$ we have

$$\|u\|_{L^p(\Gamma_g)}^p \leq 2^{p-1} \left(\frac{|\Gamma_g|}{|\Omega|} \|u\|_{L^p(\Omega)}^p + diam(\Omega)^{p-1} \ \|\nabla u\|_{L^p(\Omega)}^p \right) .$$

Proof
Choose $x_0 \in \Omega$ such that

$$u(x_0)^p \leq \left(\frac{1}{|\Omega|} \int_\Omega u \, d\sigma \right)^p \leq \frac{1}{|\Omega|} \int_\Omega u^p \, d\sigma = \frac{1}{|\Omega|} \|u\|_{L^p(\Omega)}^p \quad (2.37)$$

Again we follow the proof of Theorem 2.5 till (2.35), integrate over Γ_g, and obtain

$$\int_{\Gamma_g} u^p \, d\sigma \leq 2^{p-1} \left(|\Gamma_g| \, u(0)^p + diam(\Omega)^{p-1} \int_\Omega |\nabla u|^p \, dx \right) .$$

Now (2.37) and an extension via density to arbitrary $u \in W^{1,p}(\Omega)$ yields the assertion. □

Remark

The identification for c_{r^2} in Proposition 2.8 is not sharp. A sharp identification can be given via numerical minimazation methods computing the associated first eigenvalue

$$\lambda_p = \inf_{v \in W^{1,p}(\Omega)} \left\{ \|v\|_{W^{1,p}(\Omega)}^p : \|v\|_{L^p(\Gamma)} = 1 \right\}$$

of the problem

$$div(|\nabla u|^{p-2} \nabla u) = |u|^{p-2} u \text{ in } \Omega$$
$$|\nabla u|^{p-2} \frac{\partial u}{\partial n} = \lambda_p |u|^{p-2} u \text{ on } \Gamma$$

in specific cases. For a qualitative treatment we refer to [56], where λ_p is shown to be isolated and simple for $p \in (1, \infty)$. Let us also refer to a survey on Sobolev trace inequalities which can be found in [59].

2.4.4. An extension to contractible finite path domains

In the following we consider contractible domains $\Omega \subset \mathbb{R}^d$. We recall that Ω is contractible if it continuously retracts onto a point $x_c \in \Omega$; I.e. there exists a continuous mapping $F : \Omega \times [0, 1] \to \Omega$ such that $F(x, 0) = x$ and $F(x, 1) = x_c$ for all $x \in \Omega$, [62].

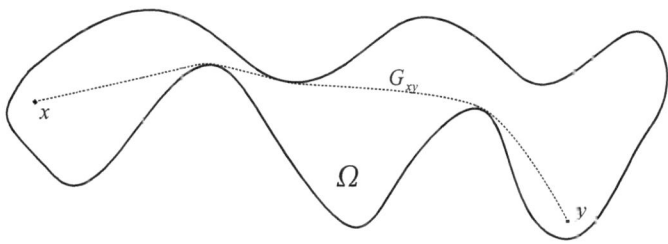

Let $\gamma_{xy} : [0,1] \to \Omega$ denote a parametrization of a piecewise differentiable geodesic path $G_{xy} \subset \Omega$ between $x, y \in \Omega$. Then, by

$$\mathrm{lgp}(\Omega) := \sup_{x,y \in \Omega} \left(\inf_{G_{xy} \subset \Omega} \int_0^1 |\dot\gamma_{xy}(s)|\, \mathrm{d}s \right)$$

we define the maximal length of a geodesic path in Ω.

Definition 2.1
$\Omega \subset \mathbb{R}^d$ is a finite path domain if it is contractible and $\mathrm{lgp}(\Omega) < \infty$.

As before, assume that the boundary Γ decomposes in Γ_β and Γ_g with $\Gamma_g \cap \Gamma_\beta = \emptyset$ and $\overline{\Gamma}_g \cup \overline{\Gamma}_\beta = \Gamma$. Thus Theorem 2.4 extends to

Proposition 2.9
Assume that $\Omega \subset \mathbb{R}^d$ is a finite path domain and $p \in [1, \infty)$. Then, for $u \in W^{1,p}(\Omega)$ we have

$$\|u\|_{L^p(\Omega)}^p \leq 2^{p-1} \left(\frac{\mathrm{lgp}(\Omega)}{d} \frac{|\Gamma|}{|\Gamma_\beta|} \|u\|_{L^p(\Gamma_\beta)}^p + \frac{\mathrm{lgp}(\Omega)^p}{p} \|\nabla u\|_{L^p(\Omega)}^p \right).$$

Proof
Analogous to the proof of Theorem 2.4 we have the estimate (2.34), assume $u \in C^1(\Omega) \cap C(\overline{\Omega})$ and $0 = x_c \in \Gamma_\beta$. For every $x \in \Omega$ we define a geodesic path $G_{0x} \subset \Omega$ which exists due to the properties of Ω. As before we obtain $u(x) - u(0) \leq \int_{G_{0x}} |\nabla u|\, \mathrm{d}\gamma$ and hence

$$u(x)^p \leq 2^{p-1} \left(u(0)^p + |G_{0x}|^{p-1} \int_{G_{0x}} |\nabla u|^p\, \mathrm{d}\gamma \right). \tag{2.38}$$

By Ω_s, $s \in (0, \text{lgp}(\Omega))$ we denote the image of a continuous retract $F : \Omega \times [0, \text{lgp}(\Omega)] \to \Omega$ to $x_c = 0$ and by Γ_s its boundary. We scale $s := \text{lgp}(\Omega_s)$ and set $|\Gamma_s| = \frac{|\Gamma|}{\text{lgp}(\Omega)^{d-1}} s^{d-1}$. $x \in \Gamma_s$, the geodesic property of G_{0x} and the scaling of s imply $|G_{0x}| \leq s$. Now an analogous proceeding as in the proof of Theorem 2.4 provides the claim. □

Corollaries 2.5 and 2.6 can be also easily extended to finite path domains.

Corollary 2.7
Assume that $\Omega \subset \mathbb{R}^d$ is a finite path domain and $p \in [1, \infty)$. Then, for $u \in W^{1,p}(\Omega)$ we have $\|u\|_{L^p(\Gamma_g)}^p \leq 2^{p-1} \left(\frac{|\Gamma_g|}{|\Gamma_\beta|} \|u\|_{L^p(\Gamma_\beta)}^p + \text{lgp}(\Omega)^{p-1} \|\nabla u\|_{L^p(\Omega)}^p \right)$
and $\|u\|_{L^p(\Gamma_g)}^p \leq 2^{p-1} \left(\frac{|\Gamma_g|}{|\Omega|} \|u\|_{L^p(\Omega)}^p + \text{lgp}(\Omega)^{p-1} \|\nabla u\|_{L^p(\Omega)}^p \right)$.

2.5. Combining of the estimates

At the end of this chapter we show that the asymptotic behaviour of the solution $u(t)$ of the full problem (2.1) does not depend on the order of the limits $t \to \infty$, $\varsigma \to 0$, $l \to \infty$. For this purpose we combine the estimates of Proposition 2.1, Proposition 2.5 and Theorem 2.3.

2.5.1. Setting of the general problem

Let us consider the cylindrical domain $\Omega_l = \Omega_{cr} \times (-l, l) \subset \mathbb{R}^d$ with the cross-sectional Lipschitz domain $\Omega_{cr} \subset \mathbb{R}^{d-1}$ for some variable length $l > 0$. To this specific geometry we formulate problem (2.1)

$$\frac{\partial u(t)}{\partial t} = \text{div}\left(\Lambda_\infty \nabla u(t)\right) + \varsigma r(\cdot, u(t)) + f_\infty \quad \text{in } \Omega_l; \ t \in (0, \infty) \quad (2.39)$$

with the initial condititon $u(0) = u_{init} \in H^1(\Omega_l)$ and the boundary conditions

$$-\left(\Lambda_\infty \nabla u(t)\right) n = \beta(u(t)) \quad \text{on } \Gamma_\beta$$
$$\left(\Lambda_\infty \nabla u(t)\right) n = g_1 \quad \text{on } \Gamma_1; \quad \left(\Lambda_\infty \nabla u(t)\right) n = g_2 \quad \text{on } \Gamma_2.$$

We have as before $\Gamma_g = \Gamma_1 \cup \Gamma_2$ with $\Gamma_1 = \Omega_{cr} \times \{-l\}$, $\Gamma_2 = \Omega_{cr} \times \{l\}$ and $\Gamma_\beta = \partial\Omega_{cr} \times [-l, l]$. The used data fulfill the requirements made for the asymptotics $t \to \infty$, $\varsigma \to 0$ and $l \to \infty$. In particular the conductivity/diffusivity matrix $\Lambda_\infty \in L^\infty(\Omega_l, \mathbb{R}^{d\times d})$ has the specific structure of (2.23) and $f_\infty \in L^2(\Omega_l)$ is constant w.r.t. the axial coordinate x_d.

Remark

In order to obtain a uniform comparison of the different asymptotics we choose the space $H^1(\Omega_{l_0})$ for some fixed $l_0 > 0$. The combining estimate will be made w.r.t. $\|\cdot\|_{L^2(\Omega_{l_0})}$ since it is the weakest norm of the involved estimates. To this end we consider the restricted solution of (2.39): $\left(u(t)|_{\Omega_{l_0}}\right)_{t\geq 0} \subset H^1(\Omega_{l_0})$, $l_0 < l$ in the following. I.e. the generalized Friedrichs constant $c_\star$ is defined w.r.t. the domain Ω_{l_0} and *not* w.r.t. Ω_l.

Existence, Uniqueness and Boundedness

Using the investigations of sections 2.1.2 and 2.1.3 suppose $|\varsigma| < \frac{\lambda_{min}}{L_r c_\star^2}$. Then there exists a unique evolution $(u(t))_{t\geq 0} \subset H^1(\Omega_{l_0})$ which solves (2.39) and converges to a stationary solution $u_\varsigma \in H^1(\Omega_{l_0})$ which is bounded by

$$\left(\lambda_{min} - L_r c_\star^2 |\varsigma|\right) \|u_\varsigma\|_{\star, l_0} \leq C_{f_\infty, g} + C_{r,\beta}$$

where $C_{f_\infty, g} = c_\star \|f_\infty\|_{L^2(\Omega_{l_0})} + C_1 \|g_1\|_{L^2(\Gamma_1)} + C_2 \|g_2\|_{L^2(\Gamma_2)}$ and $C_\beta = \sqrt{\frac{|\Gamma_\beta| \lambda_{min}}{c_\beta}} |\beta(0)|$. $C_i := \|\tau\|_{tr} = \sup_{\|v\|_{\star, l_0} \leq 1} \|\tau(v)\|_{L^2(\Gamma_i)}$ denotes the norm of the trace map $\tau : H^1(\Omega_{l_0}) \to L^2(\Gamma_i)$.

2.5.2. Setting of the reduced problem

On the other hand we consider the reduced problem

$$\begin{aligned} -\text{div}\left(\bar{\Lambda}\,\nabla\bar{u}\right) &= \bar{f} \quad \text{in } \Omega_{cr} \\ -\left(\bar{\Lambda}\,\nabla\bar{u}\right) n &= \beta(\bar{u}) \quad \text{on } \partial\Omega_{cr}. \end{aligned} \qquad (2.40)$$

where $\bar{\Lambda} \in L^\infty(\Omega_{cr}, \mathbb{R}^{(d-1)\times(d-1)})$ and $\bar{f} \in L^2(\Omega_{cr})$ are connected to Λ_∞ and f_∞ via the considerations in section 2.3.1. The existence and uniqueness of a solution $\bar{u} \in H^1(\Omega_{cr})$ of (2.40) is guaranteed by Lemma 2.4 and it is bounded by $\bar{\lambda}_{min} \|\bar{u}\|_{\star,cr} \leq c_{\star,cr} |\bar{f}\|_{L^2(\Omega_{cr})} + \sqrt{\frac{|\partial \Omega_{cr}| \bar{\lambda}_{min}}{c_\beta}} |\beta(0)|$ where $c_{\star,cr}$ denotes the generalized Friedrichs-constant of Ω_{cr}. Finally we extend $\bar{u}$ to $u_\infty \in H^1(\Omega_{l_0})$ analogously to section 2.3.1. This solution represents the limit of the solution of (2.39) w.r.t. $t \to \infty$, $\varsigma \to 0$ and $l \to \infty$.

2.5.3. Combining estimate

Now we give an estimate for the difference between the solution $u(t)_{t\geq 0} \subset H^1(\Omega_{l_0})$ and its asymptotic approximation $u_\infty \in H^1(\Omega_{l_0})$ w.r.t. $\|\cdot\|_{L^2(\Omega_{l_0})}$.

Theorem 2.5
Let $u(t)$ and u_∞ solve (2.39) and (2.40) respectively. Then there holds

$$\|u(t) - u_\infty\|_{L^2(\Omega_{l_0})} \leq C_t e^{-\phi t} + C_\varsigma |\varsigma| + C_l \exp\left(\frac{-(l-l_0)}{c_\lambda}\right) \text{ with}$$

$$\phi = \frac{\lambda_{min}}{c_\star^2} - L_r |\varsigma|, \quad c_\lambda = \frac{c_{\star,cr} \lambda_{ddmax}}{\lambda_{min}}$$

$$C_t = \|u_{init}\|_{L^2(\Omega_{l_0})} + \frac{1}{e_\star \phi} (C_{f_\infty,g} + C_{r,\beta}) \; ; \; C_\varsigma = \frac{1}{\phi} \frac{c_\star L_r (C_{f_\infty,g} + C_{r,\beta})}{\lambda_{min}}$$

$$C_l = \frac{c_\star}{\lambda_{min}} \left(C_1 \|g_1\|_{L^2(\Gamma_1)} + C_2 \|g_2\|_{L^2(\Gamma_2)} \right)$$

Proof
Let u_ς and $u_l \in H^1(\Omega_{l_0})$ denote the solutions of (2.5) and (2.20) subjected to the data of (2.39). The triangle inequality yields

$$\|u(t) - u_\infty\|_{L^2(\Omega_{l_0})} \leq \|u(t) - u_\varsigma\|_{L^2(\Omega_{l_0})} + \|u_\varsigma - u_l\|_{L^2(\Omega_{l_0})} + \|u_l - u_\infty\|_{L^2(\Omega_{l_0})}.$$

An estimate of the first summand is given by Proposition 2.1.

$$\|u(t) - u_\varsigma\|_{L^2(\Omega_{l_0})} \leq e^{-\phi t} \left(\|u_{init}\|_{L^2(\Omega_{l_0})} + \|u_\varsigma\|_{L^2(\Omega_{l_0})} \right)$$

Observe $\|u_\varsigma\|_{L^2(\Omega_0)} \leq c_\star \|u_\varsigma\|_{\star,l_0}$; then Theorem 2.1 yields a bound for $\|u_\varsigma\|_{\star,l_0}$ and we get $\|u(t) - u_\varsigma\|_{L^2(\Omega_{l_0})} \leq e^{-\phi t}\left(\|u_{init}\|_{L^2(\Omega_{l_0})} + \frac{1}{c_\star \phi}(C_{f_\infty,g} + C_{r,\beta})\right)$. The second summand $\|u_\varsigma - u_l\|_{L^2(\Omega_{l_0})} \leq c_\star \|u_\varsigma - u_l\|_{\star,l_0}$ is bounded by Proposition 2.5 $\|u_\varsigma - u_l\|_{L^2(\Omega_{l_0})} \leq \frac{|\varsigma| c_\star^2}{\lambda_{min} - |\varsigma| c_\star^2 L_r}\left(L_r \|u_l\|_{L^2(\Omega_{l_0})} + \|r(\cdot,0)\|_{L^2(\Omega_{l_0})}\right)$. This and the bound for $\|u_l\|_{L^2(\Omega_{l_0})}$ via Theorem 2.1 ($\varsigma = 0$) gives C_ς. The estimate of the last addend is given directly by Theorem 2.3. $\square$

Remark
The final combining estimate in Theorem 2.5 has a rough character due to the possibly large 'distance' between $u(t)$ and u_∞. The use of the triangle inequality and the separate results with the related detour fortify this impression. It is feasible to overcome this issue using numerical methods directly for $u(t)$, u_∞ and a posteriori estimates for an appropriate comparison.
On the other hand we are able to compare the -in this context - most simple solution u_∞ of (2.40) with the solution $u(t)$ of the full problem (2.1) directly, identifying the occuring constants explicitly. Moreover, the asymptotic behaviour of $u(t)$ towards u_∞ described in Theorem 2.5 does *not* depend on the order of the limits $t \to \infty, \varsigma \to 0$ and $l \to \infty$. Nevertheless, the choice of the sequences $\varsigma = (\varsigma_n)_{n\in\mathbb{N}} \xrightarrow[n\to\infty]{} 0$ and $l = (l_n)_{n\in\mathbb{N}} \xrightarrow[n\to\infty]{} \infty$ has to be consistent with the subresonance condition of Theorem 2.1 to guarantee the existence of the solution $u(t)$. I.e. $|\varsigma_n| < \frac{\lambda_{min}}{L_r c_\star^2}$; suppose l is large enough, such that we identify Friedrichs constant $c_\star$ with $diam(\Omega_l) = \sqrt{diam(\Omega_{cr})^2 + (2\,l)^2}$ as in the large scale case of Proposition 2.6. Then a sufficient consistency condition for $(l_n)_{n\in\mathbb{N}}$ and $(\varsigma_n)_{n\in\mathbb{N}}$ reads as $|\varsigma_n| < \frac{\lambda_{min}}{L_r(diam(\Omega_{cr})^2 + (2\,l_n)^2)}$.
The major advantage in applications is a considerable minimization of computational effort when replacing (2.1) by (2.40); taking into account an error which is controlled via Theorem 2.5. Modelling a heat transfer problem in electric cables, the next chapter will show how the constants C_t, C_ς and C_l are identified with concrete geometrical and physical quantities.

3. Estimates for heat transfer in electric cables

In many fields of modern technology it is necessary to find optimal geometric and material parameters of electric cables. For this reason, it is important to develop effective procedures that permit the direct determination of temperature at characteristic positions of the conductor.

The purpose of this chapter is the reduction of a full model problem describing dynamical heat transfer in electric cables, to a stationary, linearized cross-sectional problem. Hereto we apply the asymptotic results of chapter 2, which control the error arising when solutions of the full problem are approximated with solutions of the reduced one. The completely reduced problem is treated in chapter 4 then. In section 3.1 we consider a uninsulated cable consisting of a homogeneous material. section 3.2 deals with insulated cables.

3.1. Estimates for a uninsulated cable

3.1.1. Modelling of the heat transfer problem

The uninsulated cable is modelled as a cylindrical domain $\Omega_{l_0} = \Omega_{cr} \times (-l_0, l_0) \subset \mathbb{R}^3$ (i.e. $d = 3$) with an open cross section $\Omega_{cr} \subset \mathbb{R}^2$ and some fixed length $l_0 > 0$.
Analogously to section 2.3 we have $\Gamma_g = \Gamma_1 \cup \Gamma_2$ with $\Gamma_1 = \Omega_{cr} \times \{-l_0\}$, $\Gamma_2 = \Omega_{cr} \times \{l_0\}$ for the Neumann boundary. The monotone transfer boundary is the cylinder jacket, i.e. $\Gamma_s = \partial \Omega_{cr} \times [-l_0, l_0]$.
For the occuring physical entities we use the following notation.

I	electric current
U	voltage
ρ	electrical resistivity of the conductor material
λ	heat conductivity of the conductor material
γ	volume specific heat capacity of the conductor material
u_{env}	temperature of the environment (air)
u	temperature distribution in Ω_{l_0}
α	heat transfer coefficient on the conductor surface
$g_i;\ i=1,2$	heat flux density at the cross-sectional ends Γ_i

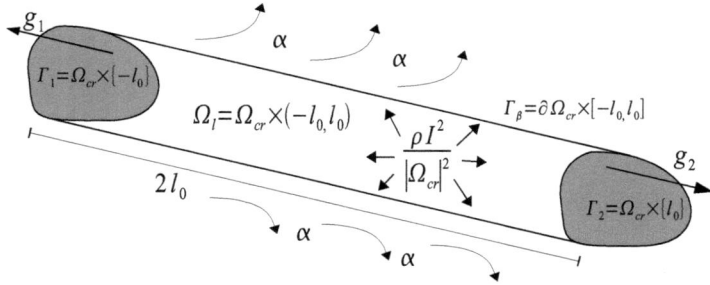

We consider the dynamical heat transfer problem

$$\gamma \frac{\partial u(t)}{\partial t} = \lambda \Delta u(t) + \frac{\rho I^2}{|\Omega_{cr}|^2} \quad \text{in } \Omega_{l_0} \tag{3.1}$$

$$-\lambda \frac{\partial u(t)}{\partial n} = \alpha(u(t))\left(u(t) - u_{env}\right) \quad \text{on } \Gamma_\beta \tag{3.2}$$

$$\lambda \frac{\partial u(t)}{\partial n} = g_1 \text{ on } \Gamma_1\ ;\ \lambda \frac{\partial u(t)}{\partial n} = g_2 \text{ on } \Gamma_2$$

with the initial condition $u(0) = u_{init}$. The structure of the source term $\frac{\rho I^2}{|\Omega_{cr}|^2}$ in (3.1) will be treated in the following paragraph. The negative sign on the left hand side of (3.2) signifies that the heat transfer $\lambda \frac{\partial u}{\partial n}$ on the surface Γ_β is directed from regions with higher temperature to regions with lower temperature; which is the Clausius statement of the second law of thermodynamics.

Derivation of the source term $\frac{\rho I^2}{|\Omega_{cr}|^2}$

Let $f_0 \in L^2(\Omega_{cr})$ denote the source term on the right hand side of (3.1). Thus $\int_{\Omega_{l_0}} f_0 \, dx$ identifies the electrical power dissipation UI in Ω_{l_0}. Initially assume that the heat power density f_0 is constant. Ohm's law and the specification of the electrical resistance yields

$$\int_{\Omega_{l_0}} f_0 \, dx = U I = \rho \frac{2 l}{|\Omega_{cr}|} I^2$$

Due to the cylindrical form of Ω_{l_0} we have $|\Omega_{l_0}| = |\Omega_{cr}| \, 2l$ and thus $f_0 = \frac{\rho I^2}{|\Omega_{cr}|^2}$. Since this argumentation can be applied to every measurable subset of Ω_{l_0}, the equation for f_0 holds also for possibly non-constant resistivities ρ.

Dependence of $\alpha = \alpha(u)$

On the right hand side of (3.2) we find the emitted sectoral heat power that involves the temperature dependent heat transfer coefficient $\alpha = \alpha(u)$. It is defined as the factor of proportionality between the emitted heat power and the difference $u - u_{env}$. Due to various fluid mechanical properties, it depends on the geometry of the heat emitting solid. For the temperature dependence of α in general we refer to [14], [18], [70], [75]. We will specify it in the case of rotational symmetry in section 3.1.5.

Specification of γ, λ **and** $\rho = \rho(u)$

Following [46] we postulate the standard model of a linear-affine temperature dependence of $\rho : \mathbb{R}^+ \to \mathbb{R}^+$ by $\rho(u) = \rho_0 \left(1 + \alpha_\rho (u - u_0)\right)$; $u = u(x)$. $\rho_0 > 0$ denotes the resistivity value to a reference temperature u_0, $\alpha_\rho \in \mathbb{R}$ identifies the linear temperature coefficient of ρ. For the sake of simplicity we set $u_0 = 0$. Assume moreover that the heat conductivity $\lambda > 0$ and the heat capacity $\gamma > 0$ is constant. These assumptions provide accurate approximations to experimental data of many conductor materials.

3.1.2. Identification of the general setting with physical quantities

In (2.1) we introduced the general data $\Lambda \in L^\infty(\Omega_{l_0}, \mathbb{R}^{3\times 3})$, $f \in L^2(\Omega_{l_0})$, $\varsigma \in \mathbb{R}$ and the continuous maps $r : \Omega_{l_0} \times \mathbb{R} \to \mathbb{R}$; $\beta : \mathbb{R} \to \mathbb{R}$. Thus we have

$$\Lambda = \frac{\lambda}{\gamma}\begin{pmatrix} 1 & 0 & 0 \\ 0 & 1 & 0 \\ 0 & 0 & 1 \end{pmatrix} \quad \text{i.e.} \quad \lambda_{min} = \frac{\lambda}{\gamma}, \quad f = \frac{\rho_0 \, I^2}{\gamma \, |\Omega_{cr}|^2}, \quad r(u) = \frac{\rho_0 \, I^2}{\gamma \, |\Omega_{cr}|^2} u$$

$$\varsigma = \alpha_\rho, \quad \beta(u) = \frac{\alpha(u)}{\gamma}(u - u_{env}), \quad g = \frac{1}{\gamma}(g_1 \, \mathbb{I}_{\{\Gamma_1\}} + g_2 \, \mathbb{I}_{\{\Gamma_2\}}).$$

Growth condition on β and Lipschitz condition on r in physical quantities
By the identification above, the Lipschitz constant for r reads as $L_r = \frac{\rho_0 \, I^2}{\gamma \, |\Omega_{cr}|^2}$, i.e. the heat power density in the conductor divided by γ.
To identify the monotonicity constant c_β and to ensure the growth condition on β, we truncate and extend the monotone and continuous heat transfer coefficient α:

$$\tilde{\alpha}(u) := \begin{cases} \alpha_l & \text{for } u < u_l \\ \alpha_h & \text{for } u > u_h \\ \alpha(u) & \text{in } [u_l, u_h] \end{cases} \tag{3.3}$$

where $0 < \alpha(u_l) = \alpha_l < \alpha(u_h) = \alpha_h$ for $u_l < u_h$.

Assume that $\beta(u) = \frac{\tilde{\alpha}(u)}{\gamma}(u - u_{env})$ is differentiable for $u \in [u_l, u_h]$ and $u_{env} \leq u_l$. Then we have

$$\inf_{s \in [u_l, u_h]} |\beta'(s)| \geq \frac{\alpha_l}{\gamma} =: c_\beta \tag{3.4}$$

The identification of c_β with the estimate $\frac{\alpha_l}{\gamma}$ is not the optimal monotonicity constant for (2.4). Nevertheless, the relation (2.4) holds.

Remark
In view of applications it makes sense to consider bounded temperature intervals. I.e. the truncation in (3.3) outside of the interval $[u_l, u_h]$ does not change the heat transfer in the relevant temperature range.

3.1.3. Subresonant states and long time behaviour

First we formulate the existence and uniqueness result for a stationary solution $u_{st} := u_\varsigma$ of (3.1) from Theorem 2.1 in the given physical setting; i.e. for

$$-\lambda \Delta u_{st} = \frac{\rho I^2}{|\Omega_{cr}|^2} \quad \text{in } \Omega_{l_0} \tag{3.5}$$

$$-\lambda \frac{\partial u_{st}}{\partial n} = \alpha \left(u_{st} - u_{env} \right) \quad \text{on } \Gamma_\beta$$

$$\lambda \frac{\partial u_{st}}{\partial n} = g_1 \text{ on } \Gamma_1 \,;\, \lambda \frac{\partial u_{st}}{\partial n} = g_2 \text{ on } \Gamma_2 \,.$$

For this purpose we define the norm on $H^1(\Omega_{l_0})$ via the identification of c_β and λ_{min}

$$\|v\|_{*,l_0}^2 = \|\nabla v\|_{L^2(\Omega_{l_0})}^2 + \frac{\alpha_l}{\lambda} \|v\|_{L^2(\Gamma_\beta)}^2 \,.$$

Corollary 3.1 (Subresonance in uninsulated cables)
Let $\alpha_\rho < \frac{\lambda |\Omega_{cr}|^2}{\rho_0 I^2 c_*^2}$. Then there exists a unique solution $u_{st} \in H^1(\Omega_{l_0})$ of (3.5) which is bounded by

$$\left(\lambda - \frac{\rho_0 I^2 |\alpha_\rho|}{|\Omega_{cr}|^2} c_*^2 \right) \|u_{st}\|_{*,l_0} \leq C_{\rho,g} + C_\alpha$$

where $C_{\rho,g} = c_* \sqrt{|\Omega_{l_0}|} \frac{\rho_0 I^2}{|\Omega_{cr}|^2} + C_1 \|g_1\|_{L^2(\Gamma_1)} + C_2 \|g_2\|_{L^2(\Gamma_2)}$
and $C_\alpha = \sqrt{\frac{|\Gamma_\beta|\lambda}{\alpha_l}} |\alpha_l u_{env}|$. $C_i := \|\tau\|_{tr} = \sup_{\|v\|_{*,l_0} \leq 1} \|\tau(v)\|_{L^2(\Gamma_i)}$ denotes the
norm of the trace map $\tau : H^1(\Omega_{l_0}) \to L^2(\Gamma_i)$.

Physical interpretation of subresonant states in uninsulated cables
First we identify the generalized Friedrichs-constant c_* for the given geomet-

rical setting of Ω_{l_0}.
There holds $diam(\Omega_{l_0})^2 = diam(\Omega_{cr})^2 + (2\,l_0)^2$, and $|\Gamma_\beta| = 2\,|\partial\Omega_{cr}|\,l_0$, $|\Gamma| = 2\,|\Omega_{cr}| + 2\,|\partial\Omega_{cr}|\,l_0$. Thus the scaling condition for the small scale case of Proposition 2.6 reads as

$$diam(\Omega_{cr})^2 + (2\,l_0)^2 \leq \frac{4\,\lambda^2}{9\,\alpha_l^2}\left(1 + \frac{|\Omega_{cr}|}{|\partial\Omega_{cr}|\,l_0}\right)^2.$$

This is fulfilled e.g. if $diam(\Omega_{cr}) \leq \frac{2\lambda}{3\alpha_l}$ and $0 < l_0^2 \leq \frac{\lambda\,|\Omega_{cr}|}{3\,\alpha_l\,|\partial\Omega_{cr}|}$. With that we identify $c_\star$ via

$$c_\star^2 = \frac{2\lambda}{3\alpha_l}\left(1 + \frac{|\Omega_{cr}|}{|\partial\Omega_{cr}|\,l_0}\right)\,diam(\Omega_{l_0}). \tag{3.6}$$

In the large scale case $diam(\Omega_{cr})^2 + (2\,l_0)^2 > \frac{4\lambda^2}{9\alpha_l^2}\left(1 + \frac{|\Omega_{cr}|}{|\partial\Omega_{cr}|\,l_0}\right)^2$ we set simply $c_\star = diam(\Omega_{l_0})$.

Now the subresonance condition of Corollary 3.1 reads as

$$|\alpha_\rho| < \frac{3}{2}\,\frac{\alpha_l\,|\Omega_{cr}|^2\,|\partial\Omega_{cr}|\,l_0}{\rho_0\,I^2\,(|\partial\Omega_{cr}|\,l_0 + |\Omega_{cr}|)\,diam(\Omega_{l_0})}$$

in the small scale case and $|\alpha_\rho| < \frac{\lambda\,|\Omega_{cr}|^2}{\rho_0\,I^2\,diam(\Omega_{l_0})^2}$ in the large scale case. This means that subresonance is given if the raising heating term $\frac{|\alpha_\rho|\,\rho_0\,I^2}{|\Omega_{cr}|^2}$ is controlled by the thermal output term $\frac{3}{2}\,\frac{\alpha_l\,|\partial\Omega_{cr}|\,l_0}{(|\partial\Omega_{cr}|\,l_0 + |\Omega_{cr}|)\,diam(\Omega_{l_0})}$ in the small scale case or $\frac{\lambda}{diam(\Omega_{l_0})^2}$ in the large scale case.

Remark
Since the resonance map $r(s) = \frac{\rho_0\,I^2}{\gamma\,|\Omega_{cr}|^2}\,s$ is monotonically increasing we need no absolute value of the temperature coefficients α_ρ in the subresonance condition of Corollary 3.1.

For materials with $\alpha_\rho < 0$ the estimate in Corollary 3.1 holds with

$$\lambda \|u_{st}\|_{\star,l_0} \leq C_{\rho,g} + C_\alpha$$

for arbitrary large $|\alpha_\rho|$. I.e. a damping term on the right hand side of (3.1) guarantees the existence of a stationary solution for any current values I.

Existence of $u(t)$ and convergence to a stationary solution u_{st}

Suppose that the subresonance condition $\alpha_\rho < \frac{\lambda |\Omega_{cr}|^2}{\rho_0 I^2 c_\star^2}$ from Corollary 3.1 holds. Then there exists a unique solution $u \in L^1([0,\infty), H^1(\Omega_{l_0}))$ of (3.1) via Theorem 2.2. Moreover, by Proposition 2.1, we have an exponential rate of convergence of $(u(t))_{t\in[0,\infty)}$ to the stationary solution u_{st} of (3.5).

Corollary 3.2

Let $u(t)$ and u_{st} denote the solutions of (3.1) and (3.5) respectively and let $\alpha_\rho < \frac{\lambda |\Omega_{cr}|^2}{\rho_0 I^2 c_\star^2}$ hold. Then we have

$$\|u(t) - u_{st}\|_{L^2(\Omega_{l_0})} \leq e^{-\phi t} \|u_{init} - u_{st}\|_{L^2(\Omega_{l_0})} \quad \text{where } \phi = \frac{1}{\gamma}\left(\frac{\lambda}{c_\star^2} - \frac{\rho_0 |\alpha_\rho| I^2}{|\Omega_{cr}|^2}\right).$$

To obtain an expression for ϕ which depends on the physical parameters of the cable only, we can identify $c_\star = diam(\Omega_{l_0})$ in the large scale case or via (3.6) in the small scale case.

Remark
The estimate in Corollary 3.2 can be improved for any negative temperature coefficient α_ρ by $\phi = \frac{\lambda}{\gamma c_\star^2}$. This means that the temperature damping effect of negative α_ρ causes a faster convergence of $u(t)$ towards u_{st}.

Investigation of constant temperature profiles

Suppose now that the initial boundary value problem in (3.1), (3.2) has homogeneous Neumann data $g_i = 0$, $i = 1, 2$. Following section 2.1.4 we introduce an energy conservating mean value $(u^{mv}(t))_{t\in[0,\infty)} \subset \mathbb{R}$ of the solution of (3.1) which is constant in space. It solves the ordinary differential equation

$$\dot{u}^{mv} = \frac{\rho_0 \, I^2}{\gamma \, |\Omega_{cr}|^2} \left(1 + \alpha_\rho \, u^{mv}\right) - \frac{|\partial\Omega_{cr}|}{\gamma \, |\Omega_{cr}|} \, \tilde{\alpha}(u^{mv}) \, (u^{mv} - u_{env}) \quad (3.7)$$
$$u^{mv}(0) = u^{mv}_{init}$$

Due to the truncation of α in (3.3) the right hand side of (3.7) is globally Lipschitz continuous with respect to u^{mv}. I.e. there exists a unique solution of (3.7) in $(0, \infty)$ via the Picard Lindelöf Theorem.

Moreover, by Corollary 2.1 the mean value evolution $u^{mv} = u^{mv}(t)$ converges to a stationary solution $u^{mv}_{st} \in \mathbb{R}$ of (3.7), i.e. of

$$\frac{\rho_0 \, I^2}{|\Omega_{cr}|} \left(1 + \alpha_\rho \, u^{mv}_{st}\right) = |\partial\Omega_{cr}| \, \tilde{\alpha}(u^{mv}_{st}) \, (u^{mv}_{st} - u_{env}) \quad (3.8)$$

if the relation $\frac{\alpha_\rho \rho_0 I^2}{|\Omega_{cr}|} < |\partial\Omega_{cr}| \alpha_l$ holds.

It remains to give an estimate for the rate of convergence of u^{mv} to u^{mv}_{st}. Since r fulfills the monotonicity estimate (2.15) with $c_r = L_r = \frac{\rho_0 I^2}{\gamma |\Omega_{cr}|^2}$ we can use the improved estimate for possibly negative α_ρ of Corollary 2.2 and obtain

$$|u^{mv}(t) - u^{mv}_{st}| \le e^{-\phi^{mv} t} \, |u^{mv}_{init} - u^{mv}_{st}| \; ; \; \phi^{mv} := \frac{1}{\gamma |\Omega_{cr}|} \left(|\partial\Omega_{cr}| \alpha_l - \frac{\rho_0 \alpha_\rho I^2}{|\Omega_{cr}|}\right).$$

Remark

If α is not truncated by (3.3) the equation (3.8) posesses a solution for arbitrary large values of α_ρ; I.e. no resonance effect occurs. If $u = u(t)$ is constant in space (e.g. u^{mv}), the generated heat is immediately transported to the boundary of Ω. There we have a superlinear growth of (the natural) α with $\alpha(u) \sim u^3$ due to the Stefan-Boltzmann-law for radiative heat transfer, [75]. I.e. we get the existence of a thermodynamical equilibrium for every $\alpha_\rho \in \mathbb{R}$. This is in contrast to Corollary 3.1 where we have a non-constant temperature profile, i.e. a finite heat conductivity λ. Hence a thermal resistance in Ω causes a temperature evolution towards infinity for large values of ς. This situation and a sufficient subresonance condition for α_ρ is given in Corollary 3.1. Hence, the truncation of α in (3.3) makes sense. The possible equilibria in (3.8) for natural α can yield temperatures of a magnitude where the modeling of the physical situation in (3.1) as well as the assumption of a constant temperature profile u^{mv} is not adequate anymore.

We will use the explicit euler scheme from (2.16) for determination of solutions of (3.7) in section 3.1.5 applying it to physical data.

Exponential growth for the case $\alpha_\rho \geq \frac{\lambda |\Omega_{cr}|^2}{\rho_0 I^2 c_*^2}$

In this case the asymptotic behaviour of possible solutions u of (3.1) is unclear. Analogous to Propositition 2 we establish an exponential growth estimate for $u = u(t)$, $t \in (0, t_{max})$ supposing α_ρ is large enough. To this end we consider again the initial boundary value problem (3.1) with idealized boundary conditions, i.e.

$$\gamma \frac{\partial u(t)}{\partial t} = \lambda \Delta u(t) + \frac{\rho_0 (1 + \alpha_\rho u) I^2}{|\Omega_{cr}|^2} \quad \text{in } \Omega_{l_0} \tag{3.9}$$

$$-\lambda \frac{\partial u(t)}{\partial n} = \tilde{\alpha}(u(t))(u(t) - u_{env}) \quad \text{on } \Gamma_\beta \, ; \quad -\lambda \frac{\partial u(t)}{\partial n} = 0 \quad \text{on } \Gamma_g$$

and $u(0) = u_{init} \in H^1(\Omega_{l_0})$.

Here we identify the general data Λ, ς and β as in section 3.1.2. and redefine $f = 0$; $r(u) = \frac{\rho_0 \left(\frac{1}{\alpha_\rho} + u\right) I^2}{\gamma |\Omega_{cr}|^2}$. Hence the truncated boundary transfer map $\beta(u) = \frac{\tilde{\alpha}(u)}{\gamma}(u - u_{env})$ fulfills a sublinear growth condition with $L_\beta = \frac{\alpha_h}{\gamma}$. The resonance map $r = r(u)$ satisfies a superlinear growth condition with

$$r(u) \geq \frac{\rho_0 I^2}{\gamma |\Omega_{cr}|^2} u \quad \Longrightarrow \quad r_{min} = \frac{\rho_0 I^2}{\gamma |\Omega_{cr}|^2}.$$

Again we consider the energy conserving mean value $(u^{nv}(t))_{t \in [0, t_{max}]}$ of a solution of (3.9) which is constant in space. With the identifications for β and r and by (2.19) we obtain the initial value problem in (3.7). Assume now that there exists a solution of (3.7) in $[0, t_{max}]$.

Corollary 3.3
Let u^{mv} denote a solution of (3.7) and let $\alpha_\rho \geq \frac{\alpha_h |\partial\Omega_{cr}| |\Omega_{cr}|}{\rho_0 I^2}$, then there holds

$$|u^{mv}(t)| \geq |u^{mv}_{init}| e^{\phi_{res} t} \quad \text{where} \quad \phi_{res} := \frac{1}{\gamma |\Omega_{cr}|} \left(\frac{\rho_0 \alpha_\rho I^2}{|\Omega_{cr}|} - |\partial\Omega_{cr}| \alpha_h \right).$$

Note that the exponential growth condition $\alpha_\rho \geq \frac{|\Gamma|}{2 l_0} \frac{\alpha_h |\Omega_{cr}|}{\rho_0 I^2}$ plausibly implies the condition $\alpha_\rho \geq \frac{\lambda |\Omega_{cr}|^2}{\rho_0 I^2 c_\star^2}$ for possible resonance; provided the inequality $diam(\Omega_{cr}) \alpha_l < 16/9 \, \lambda$ holds in the large scale and $\alpha_h \geq \alpha_l$ in the small scale case. The last inequalities are true for any realistic setting.

3.1.4. Sensitivity for $\alpha_\rho \to 0$ and asymptotics for $l \to \infty$

Helmholtz-to-Poisson estimate in (3.5) via $\alpha_\rho \to 0$

Consider (3.5) for $\alpha_\rho = 0$, i.e.

$$\begin{aligned}
-\lambda \Delta u &= \frac{\rho_0 I^2}{|\Omega_{cr}|^2} \quad \text{in } \Omega_{l_0} & (3.10) \\
-\lambda \frac{\partial u}{\partial n} &= \alpha (u - u_{env}) \quad \text{on } \Gamma_\beta \\
\lambda \frac{\partial u}{\partial n} &= g_1 \text{ on } \Gamma_1 \; ; \; \lambda \frac{\partial u}{\partial n} = g_2 \text{ on } \Gamma_2.
\end{aligned}$$

By Corollary 3.1 there exists a unique solution $u \in H^1(\Omega_{l_0})$ of (3.10) which is bounded by $\lambda \|u\|_{\star,l_0} \leq C_{\rho,g} + C_\alpha$. We investigate the sensitivity $u_{st} \xrightarrow[\alpha_\rho \to 0]{} u$ w.r.t. $\|\cdot\|_{\star,l_0}$. Proposition 2.5 and the identifications in section 3.1.2 provide

Corollary 3.4
Assume $\alpha_\rho < \frac{\lambda |\Omega_{cr}|^2}{\rho_0 I^2 c_\star^2}$. Then the following estimate holds

$$\|u_{st} - u\|_{\star,l_0} \leq \underbrace{\frac{|\alpha_\rho| \rho_0 I^2 c_\star}{\lambda |\Omega_{cr}|^2 - |\alpha_\rho| \rho_0 I^2 c_\star^2}}_{=:C_{\alpha_\rho}} \|u\|_{L^2(\Omega_{l_0})}.$$

Remarks

(i) Observe that the heat capacity γ plausibly does not influence the estimate in Corollary 3.4 nor the forthcoming one for $l \to \infty$.

(ii) For negative α_ρ the estimate in Corollary 3.4 reads as

$$\|u_{st} - u\|_{*,l_0} \leq \frac{|\alpha_\rho|\,\rho_0\,I^2\,c_\star}{\lambda\,|\Omega_{cr}|^2}\,\|u\|_{L^2(\Omega_{l_0})}.$$

(iii) *Error minimizing choice of the Poisson-Datum in (3.10)*
Analogous to section 2.2.2 we set for the right hand side of (3.10)
$f_{\alpha_\rho} = \frac{\rho_0\,I^2\,(1+\alpha_\rho\,\bar{u})}{|\Omega_{cr}|^2}$ for some constant $\bar{u} \in \mathbb{R}$. The associated solution u of (3.10) yields the estimate $\|u_{st} - u\|_{*,l_0} \leq C_{c_\rho}\,\|u - \bar{u}\|_{L^2(\Omega_{l_0})}$ and thus a possible decrease of the error for a suitable choice of $\bar{u} \in \mathbb{R}$.

Reduction to a cross-sectional problem via $l \to \infty$

Setting $\bar{\Lambda} = \frac{\lambda}{\gamma}\begin{pmatrix} 1 & 0 \\ 0 & 1 \end{pmatrix}$ and $\bar{f} = \frac{\rho_0\,I^2}{\gamma|\Omega_{cr}|^2}$ in (2.22), we consider the cross-sectional boundary value problem

$$-\lambda\,\Delta\bar{u} = \frac{\rho_0\,I^2}{|\Omega_{cr}|^2} \quad \text{in } \Omega_{cr} \qquad (3.11)$$

$$-\lambda\,\frac{\partial\bar{u}}{\partial n} = \alpha(\bar{u})\,(\bar{u} - u_{env}) \quad \text{on } \partial\Omega_{cr}.$$

By Lemma 2.4 there exists a unique solution $\bar{u} \in H^1(\Omega_{cr})$ of (3.11) which is bounded by $\lambda\|\bar{u}\|_{*,cr} \leq \frac{c_{\star,cr}\,\rho_0\,I^2}{|\Omega_{cr}|^{3/2}} + \sqrt{\frac{|\partial\Omega_{cr}|\lambda}{\alpha_l}}\,|\alpha_l\,u_{env}|$. Here $c_{\star,cr}$ denotes the generalized Friedrichs-constant of Ω_{cr}.

The extension of the cross-sectional data to $\Omega_l \subset \mathbb{R}^3$ is simply

$$f_\infty = \bar{f} = \frac{\rho_0\,I^2}{\gamma\,|\Omega_{cr}|^2}, \quad \Lambda_\infty = \frac{\lambda}{\gamma}\begin{pmatrix} 1 & 0 & 0 \\ 0 & 1 & 0 \\ 0 & 0 & 1 \end{pmatrix}, \quad \text{i.e. } \lambda_{min} = \lambda_{ddmax} = \frac{\lambda}{\gamma}$$

and $u_\infty(x_1, x_2, x_3) = \bar{u}(x_1, x_2)$. The associated cylinder boundary value problem reads as (3.10) w.r.t. Ω_l. It remains to show the convergence of solutions of (3.10) - now labeled $(u_l)_{l>0}$ - towards the extended solution u_∞ of the cross-sectional problem (3.11) for large l.

Corollary 3.5
Let u_l denote the solution of (3.10) and u_∞ the extended solution of (3.11). Then, for $l_0 < l$ there holds

$$\lambda \|u_l - u_\infty\|_{\star, l_0} \leq \exp\left(\frac{-(l-l_0)}{c_{\star, cr}}\right) \left(C_1 \|g_1\|_{L^2(\Gamma_1)} + C_2 \|g_2\|_{L^2(\Gamma_2)}\right).$$

$C_i := \|\tau\|_{tr} = \sup\limits_{\|v\|_{\star,l_0} \leq 1} \|\tau(v)\|_{L^2(\Gamma_i)}$ denotes the norm of the trace map τ : $H^1(\Omega_{l_0}) \to L^2(\Gamma_i)$.

Remark
In the small scale case $diam(\Omega_{cr}) \leq \frac{\lambda}{\alpha_l}$ we identify $c_{\star, cr}$ via Proposition 2.6 with $c_{\star, cr} = \sqrt{\frac{\lambda}{\alpha_l} diam(\Omega_{cr})}$ and in the large scale case $diam(\Omega_{cr}) > \frac{\lambda}{\alpha_l}$ with $c_{\star, cr} = diam(\Omega_{cr})$.

3.1.5. Application to physical data

In the following all quantities are listed in corresponding SI units. For the sake of clearness let us assume homogeneous Neumann boundary conditions $g_1 = g_2 = 0$ in (3.1). I.e. - by Corollary 3.5 - we have a constant temperature profile in the axial direction. Thus we set shortly a cross-sectional problem with $\Omega = \Omega_{cr} = \{x \in \mathbb{R}^2, |x| < 0.002\}$, i.e. $d_\Omega := diam(\Omega) = 0.004$ and $\Gamma = \Gamma_\beta = \partial \Omega$.

We fix the physical data with:
$\lambda = 400$, $\rho_0 = 1.72 * 10^{-8}$, $\alpha_\rho = 3.83 * 10^{-3}$, $\gamma = 1 * 10^5$, $u_{env} = 25$.
We do not fix the current I for the moment, since it is the characteristic variable to distinguish between the subresonant and the possibly resonant state

in Corollary 3.1. It remains to specify α on cylindrical surfaces.

Specification of α in the case of rotational symmetry

We follow fluid mechanical considerations in [5], [15], [17] concerning the heat transfer coefficient α on cylindrical surfaces. Accordingly we have

$$\alpha(u) = \underbrace{\left(\frac{\alpha_d}{\sqrt{d_\Omega}} + \alpha_u \sqrt[6]{u - u_{env}}\right)^2}_{=\alpha_c} + \underbrace{\epsilon\sigma\left(\bar{u}^2 + u_{env}^2\right)(\bar{u} + u_{env})}_{=\alpha_r}.$$

Thus α decomposes in a convection part α_c and a radiation part α_r. Here σ and ϵ denote the Stefan-Boltzmann constant, respectively the emissivity of the conductor surface. The values are fixed with $\bar{u} = u_{abs} + u$ where $u_{abs} \approx 273.15K$ denotes the difference from 0°C to absolute zero and $\sigma = 5.67 * 10^{-8}$; $\epsilon = 0.06$. The parameters α_d and α_u describe the dependence of the convection part on the diameter d_Ω and the difference in temperature, respectively. They also depend on temperature, since the fluid-mechanical values of air are temperature dependent. The following figure illustrates the monotone character of $\alpha = \alpha(u)$ for several conductor diameters.

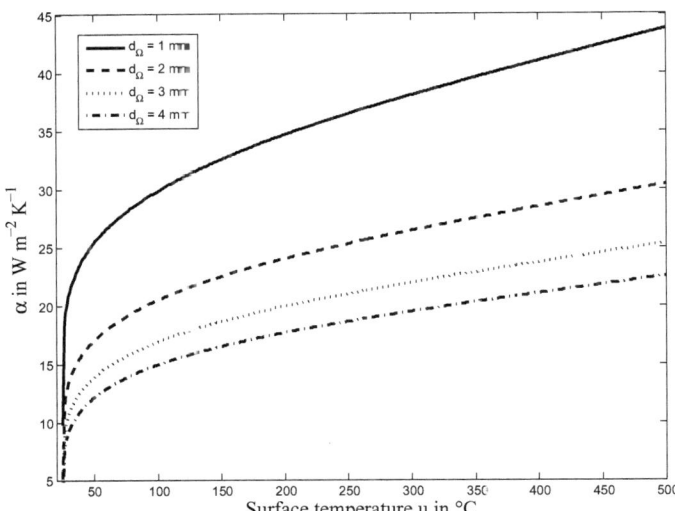

Hereby we truncate α via (3.3) at the temperatures $u_l = u_{env} = 25$ and $u_h = 500$ with $\alpha_l = 10$ and $\alpha_h = 22,5$.

Evaluation of the asymptotics for $t \to \infty$ and $\alpha_\rho \to 0$

First we observe that with the given data there holds $diam(\Omega) \leq \frac{\lambda}{\alpha_l}$, i.e. we are in the small scale case and thus $c_\star = \sqrt{\frac{\lambda}{\alpha_l}} diam(\Omega) = 0,4$. Hence Corollaries 3.1 and 3.2 read as

Let $I < 77.4$. Then there exists a unique solution $u_{st} \in H^1(\Omega)$ of (3.5) which is bounded by $\|u_{st}\|_\star \leq 1,45$ (for $I = 40$). Let $u(t)$ denote the solution of (3.1). Then we have

$$\|u(t) - u_{st}\|_{L^2(\Omega)} \leq e^{-\phi t} \|u_{init} - u_{st}\|_{L^2(\Omega)} \quad \text{where } \phi = 0,018 \ (\text{for } I = 40).$$

Finally we consider the solution of (3.5) for $\alpha_\rho \to 0$.

Let u_{st} and u denote the solutions of (3.5) and (3.10) respectively and assume $I < 77.4$ hold. Then the following estimate holds

$$\|u_{st} - u\|_\star \leq 0,04 \|u\|_{L^2(\Omega)} \quad (\text{for } I = 10).$$

Evaluation of constant temperature profiles

We evaluated the estimates for general temperature profiles $u \in H^1(\Omega)$ in the previous paragraph. This leads to restrictive subresonance conditions (I<77,4) which are far from being neccesary for the convergence of solutions $u(t)$ of (3.1) to stationary solutions u_{st} of (3.5). One reason is the possibly too rough estimate for the generalized Friedrichs constant $c_\star$. This problem can be overcome introducing an energy conservating mean value $(u^{mv}(t))_{t \in [0,\infty)} \subset \mathbb{R}$ solving (3.7) which is constant in space.

Now the respective subresonance condition - i.e. existence and uniqueness of solutions of (3.8) - is given by $I < 154,8$. In this case the solution $u^{mv}(t)$ of

(3.7) converges to the stationary solution u_{st}^{mv} of (3.8) with the following rate

$$|u^{mv}(t) - u_{st}^{mv}| \leq e^{-\phi^{mv} t} |u_{init}^{mv} - u_{st}^{mv}| \; ; \; \phi^{mv} := 0.093 \; (for \; I = 40).$$

Now we illustrate the convergence of $u^{mv} = u^{mv}(t)$ to the stationary solution u_{st}^{mv} for $I = 40$ via the Euler scheme presented in section 2.1.4. Moreover we compare the evolution $u^{mv}(t)$ with the approximating interpolation $u_{itpl}^{mv}(t) := e^{-\phi^{mv} t} u_{init}^{mv} + \left(1 - e^{-\phi^{mv} t}\right) u_{st}^{mv}$ also introduced in 2.1.4. Hereby we set $u_{init}^{mv} = u_{env} = 25$.

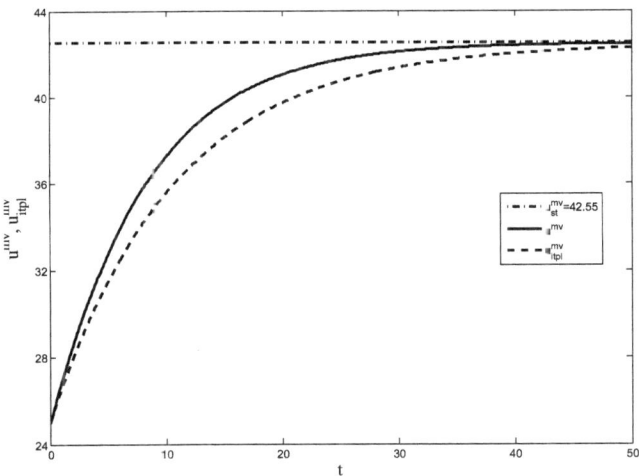

The exponential growth estimate from Corollary 3.3 reads as

Let u^{mv} denote a solution of (3.7) and let $I > 232, 4$, then there holds

$$|u^{mv}(t)| \geq |u_{init}^{mv}| e^{\phi_{res} t} \quad where \quad \phi_{res} := 0.15 \; (for \; I = 300).$$

Note that this estimate holds for $u^{mv}(t) < 500$ only since this is the upper bound for the truncation of α. The following figure shows the exponential growth of u^{mv} for $I = 300$ for truncated and natural α.

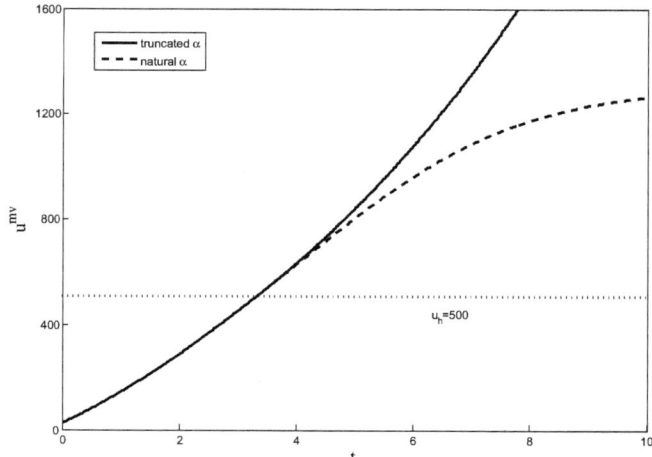

Observe that the truncation of α does not change the temperature evolution in the relevant range $10 \leq u^{mv} \leq 500$ as remarked in section 3.1.2.

3.1.6. Oscillating behaviour of stationary solutions for large temperature coefficients α_ρ

Let us illustrate the oscillating behaviour of stationary solutions of (3.5); and thus the notion of resonance in Theorem 2.1 and Corollary 3.1.

Oscillating behaviour in unbounded domains

To this end we neglect the monotone boundary condition [(3.2)] and investigate stationary solutions of (3.1) on the whole $\mathbb{R}^2$. I.e. of

$$-\lambda \Delta u = \frac{\rho_0 \left(1 + \alpha_\rho u\right) I^2}{|\Omega_{cr}|^2} \quad \text{in } \mathbb{R}^2 \tag{3.12}$$

We fix $I = 10$. Except for the temperature coefficient α_ρ, the remaining quantities are set as in the beginning of section 3.1.5.
The following diagramms show the profiles of rotationally symmetric solutions

of (3.12) in $\mathbb{R}^2$ for different values of α_p. The solutions are normed via $u(0) = 1$ and $\nabla u(0) = 0$; r denotes the distance to the origin.

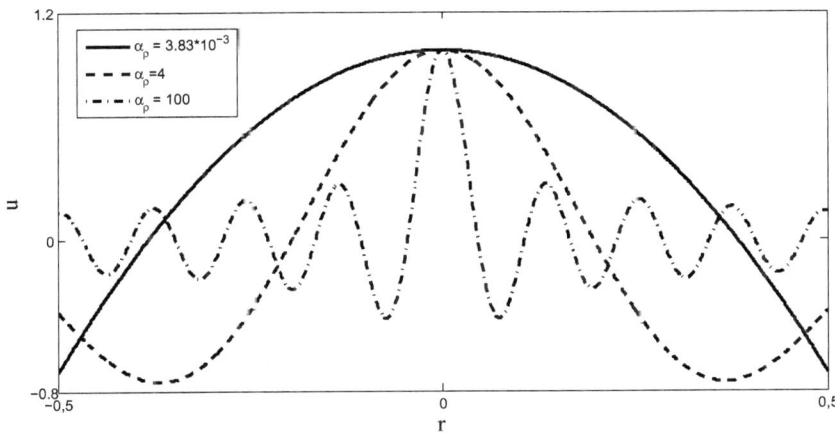

The solutions are given by Bessel functions of the first kind. They solve the ordinary differential equation which results from the rotationally symmetric transformation of the Laplace operator in (3.12).

Oscillating behaviour for large diameters

On the other hand it is possible to recover the oscillatory behaviour for the original stationary problem of (3.5). I.e. we consider

$$-\lambda \Delta u = \frac{\rho_0 (1 + \alpha_p u) I^2}{|B_r|^2} \quad \text{in } B_r(0) \subset \mathbb{R}^2 \quad (3.13)$$

$$-\lambda \frac{\partial u}{\partial n} = \alpha (u - u_{env}) \quad \text{on } \partial B_r(0)$$

with modified parameters; in particular with a large diameter $diam(B_r) = 2r$ and a large temperature coefficient α_p. The following graphics shows the solution of (3.13) for the follwing parameters.

$\lambda = 1$, $r = 1$, $\rho_0 = 1.72 * 10^{-8}$, $\alpha_p = 1 * 10^5$, $I = 1 * 10^3$, $u_{env} = 25$.

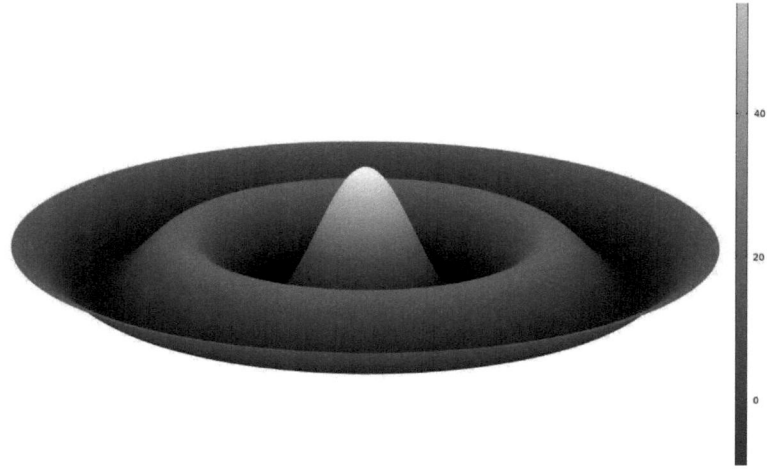

These values are distinctly beyond the subresonant state described by Corollary 3.1. I.e. the according solution has no proper physical interpretation since it is not the time limit of a corresponding dynamical problem. Nevertheless it shows the possibly oscillatory behaviour of solutions of (3.5) in large domains for large α_ρ. See e.g. [69] for existence and uniqueness results.

3.2. Estimates for an insulated cable

In the following we use the estimates of chapter 2 for insulated cables. Being important in applications, we have the difficulty of inhomogeneous material parameters here. The reduced problem for the insulated cable will be the basis boundary value problem for chapter 4.

3.2.1. Modelling of the problem

In addition to the previous notation we distinguish between the heat conductivity λ_1 for the conductor material and the heat conductivity λ_2 for the insulator material. The same indication holds for the heat capacities γ_1 and γ_2. The forthcoming sketch shows the cross section of an insulated cable.

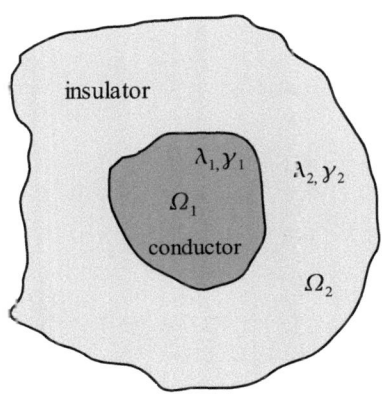

We describe the cross section of the main by the simply connected and open union $\Omega_{cr} = \overline{\Omega}_1 \cup \Omega_2 \subset \mathbb{R}^2$ with Lipschitz boundaries $\partial \Omega_{cr}$, $\partial \Omega_1$ and consider the cylindrical domain $\Omega_{l_0} = \Omega_{cr} \times (-l_0, l_0) \subset \mathbb{R}^3$, $l_0 > 0$ with the following boundary division $\Gamma_\beta = \partial\Omega_{cr} \times (-l_0, l_0)$, $\Gamma_{N_1} = \Omega_{cr} \times \{-l_0\}$ and $\Gamma_{N_2} = \Omega_{cr} \times \{l_0\}$. Thus we consider

$$\gamma_1 \frac{\partial u(t)}{\partial t} = \lambda_1 \Delta u(t) + \frac{\rho_0 (1 + \alpha_\rho u(t)) I^2}{|\Omega_1|^2} \quad \text{in } \Omega_1 \times (-l_0, l_0) \quad (3.14)$$

$$\gamma_2 \frac{\partial u(t)}{\partial t} = \lambda_2 \Delta u(t) \quad \text{in } \Omega_2 \times (-l_0, l_0)$$

$$-\lambda_2 \frac{\partial u(t)}{\partial n} = \alpha(u(t))(u(t) - u_{env}) \quad \text{on } \Gamma_\beta$$

$$\lambda_1 \frac{\partial u(t)}{\partial n} = g_{N_1 j,1} \text{ on } \Gamma_{N_j} \cap \Omega_1 \; ; \; \lambda_2 \frac{\partial u(t)}{\partial n} = g_{N_j,2} \text{ on } \Gamma_{N_j} \cap \Omega_2$$

$j = 1, 2$ and $u(0) = u_{init}$.

Identification of the general setting

The evolution of the temperature distribution $u = u(t)$ modelled by (3.14) satisfies the initial boundary value problem (2.1)

$$\frac{\partial u(t)}{\partial t} = \text{div}(\Lambda \nabla u(t)) + \varsigma r(\cdot, u(t)) + \overset{x}{\jmath} \quad \text{in } \Omega_{l_0}$$

$$-(\Lambda \nabla u(t)) n = \beta(u(t)) \quad \text{on } \Gamma_\beta$$

$$(\Lambda \nabla u(t)) n = g_1 \text{ on } \Gamma_{N_1} \; ; \; (\Lambda \nabla u(t)) = g_2 \text{ on } \Gamma_{N_2}.$$

with the following identifications.

$$\Lambda = \Lambda(x) = \frac{\lambda_1}{\gamma_1}\begin{pmatrix} 1 & 0 & 0 \\ 0 & 1 & 0 \\ 0 & 0 & 1 \end{pmatrix}\mathbb{I}_{\{\Omega_1\}}(x) + \frac{\lambda_2}{\gamma_2}\begin{pmatrix} 1 & 0 & 0 \\ 0 & 1 & 0 \\ 0 & 0 & 1 \end{pmatrix}\mathbb{I}_{\{\Omega_2\}}(x)\,,\ x \in \Omega_{l_0}$$

$$f = f(x) = \frac{\rho_0\,I^2}{\gamma_1\,|\Omega_1|^2}\mathbb{I}_{\{\Omega_1\}}(x)\,,\ r = r(x,u) = \frac{\rho_0\,I^2}{\gamma_1\,|\Omega_1|^2}\,u\,\mathbb{I}_{\{\Omega_1\}}(x)\,,\ \varsigma = \alpha_\rho$$

and $\beta(u) = \frac{\alpha(u)}{\gamma_2}(u - u_{env})$. The neumann boundary data read as

$$\begin{aligned} g_1 &= \frac{g_{N_1,1}}{\gamma_1}\mathbb{I}_{\{\Gamma_{N_1}\cap\Omega_1\}}(x) + \frac{g_{N_1,2}}{\gamma_2}\mathbb{I}_{\{\Gamma_{N_1}\cap\Omega_2\}} \\ g_2 &= \frac{g_{N_2,1}}{\gamma_1}\mathbb{I}_{\{\Gamma_{N_2}\cap\Omega_1\}}(x) + \frac{g_{N_2,2}}{\gamma_2}\mathbb{I}_{\{\Gamma_{N_2}\cap\Omega_2\}}\,,\ x \in \Omega_{l_0} \end{aligned}$$

Due to the material properties of the insulator and the conductor we have $\frac{\lambda_2}{\gamma_2} \leq \frac{\lambda_1}{\gamma_1}$ and thus $\lambda_{min} = \frac{\lambda_2}{\gamma_2}$. Moreover, by analogy to section 3.1.2 we have $L_r = \frac{\rho_0\,I^2}{\gamma_1\,|\Omega_1|^2}$ and $c_\beta = \frac{\alpha_l}{\gamma_2}$.

3.2.2. Subresonant states and long-time behaviour

The assertion of Theorem 2.1 providing a sufficient condition for the subresonant state of the stationary problem

$$\begin{aligned} -\lambda_1\,\Delta u_{st} &= \frac{\rho_0\,(1 + \alpha_\rho\,u_{st})\,I^2}{|\Omega_1|^2} \quad \text{in } \Omega_1 \times (-l_0, l_0) & (3.15) \\ -\lambda_2\,\Delta u_{st} &= 0 \quad \text{in } \Omega_2 \times (-l_0, l_0) \\ -\lambda_2\,\frac{\partial u_{st}}{\partial n} &= \alpha(u_{st})\,(u_{st} - u_{env}) \quad \text{on } \Gamma_\beta \\ \lambda_1\,\frac{\partial u_{st}}{\partial n} &= g_{N_1,1} \text{ on } \Gamma_{N_1} \cap \Omega_1\ ;\ \lambda_2\,\frac{\partial u_{st}}{\partial n} = g_{N_1,2} \text{ on } \Gamma_{N_1} \cap \Omega_2 \\ \lambda_1\,\frac{\partial u_{st}}{\partial n} &= g_{N_2,1} \text{ on } \Gamma_{N_2} \cap \Omega_1\ ;\ \lambda_2\,\frac{\partial u_{st}}{\partial n} = g_{N_2,2} \text{ on } \Gamma_{N_2} \cap \Omega_2\,. \end{aligned}$$

is given by

Corollary 3.6

Let $\alpha_\rho < \frac{\gamma_1 \lambda_2 |\Omega_1|^2}{\gamma_2 \rho_0 I^2 c_\star^2}$. Then there exists a unique stationary solution $u_{st} \in H^1(\Omega_{l_0})$ of (3.15) which is bounded by

$$\left(\frac{\lambda_2}{\gamma_2} - \frac{\rho_0 I^2 |\alpha_\rho|}{\gamma_1 |\Omega_1|^2} c_\star^2\right) \|u_{st}\|_{\star,l_0} \leq C_{\rho,g} + C_\alpha$$

where $C_{\rho,g} = c_\star \sqrt{|\Omega_{l_0}|} \frac{\alpha_0 I^2}{\gamma_1 |\Omega_1|^2} + C_1 \|g_1\|_{L^2(\Gamma_{N_1})} + C_2 \|g_2\|_{L^2(\Gamma_{N_2})}$
and $C_\alpha = \sqrt{\frac{|\Gamma_\beta| \lambda_2}{\alpha_l}} \left|\frac{\alpha_l}{\gamma_2} u_{env}\right|$. $C_i := \|\tau\|_{tr} = \sup_{\|v\|_{\star,l_0} \leq 1} \|\tau(v)\|_{L^2(\Gamma_{N_i})}$ denotes the norm of the trace map $\tau : H^1(\Omega_{l_0}) \to L^2(\Gamma_{N_i})$.

Note that the heat capacities γ_1, γ_2 influence the estimate in Corollary 3.6 eventhough we consider a stationary problem. The reason is the dynamical identification of Λ and r to have consistent interpretation of the solution of (3.15) as a limit of the solution of (3.14) for $t \to \infty$. Here we have different heat capacities in the general minimal bound on Λ which is $\frac{\lambda_2}{\gamma_2}$ and the source term $r = \frac{\rho_0 \, r^2}{\gamma_1 |\Omega_1|^2} u \, \mathbb{I}_{\{\Omega_1\}}$. They do not cancel such as in Corollary 3.1. Concerning just the stationary problem in (3.15) it is not necessary to identify Λ and r via the dynamical setting. A suggestion for the treatment of the stationary situation is given in section 3.2.3.

The convergence of the dynamical solution in (3.14) to the stationary solution of (3.15) reads as

Corollary 3.7

Let $u(t)$ and u_{st} denote the solutions of (3.14) and (3.15) respectively and let $\alpha_\rho < \frac{\gamma_1 \lambda_2 |\Omega_1|^2}{\gamma_2 \rho_0 I^2 c_\star^2}$ hold. Then we have

$$\|u(t) - u_{st}\|_{L^2(\Omega_{l_0})} \leq e^{-\frac{\phi}{2} t} \|u_{init} - u_{st}\|_{L^2(\Omega_{l_0})} \quad \text{where} \quad \phi = \frac{\lambda_2}{\gamma_2 c_\star^2} - \frac{\rho_0 |\alpha_\rho| I^2}{\gamma_1 |\Omega_1|^2}.$$

Limitations of Theorem 2.1

Corollary 3.6 gives a sufficient condition for subresonance which is very restrictive. Moreover the associated dynamical behaviour of the dynamical solution in Corollary 3.7 is very rough. This is due to the general minimal bound for

$$\Lambda = \frac{\lambda_1}{\gamma_1} \begin{pmatrix} 1 & 0 & 0 \\ 0 & 1 & 0 \\ 0 & 0 & 1 \end{pmatrix} \mathbb{I}_{\{\Omega_1\}} + \frac{\lambda_2}{\gamma_2} \begin{pmatrix} 1 & 0 & 0 \\ 0 & 1 & 0 \\ 0 & 0 & 1 \end{pmatrix} \mathbb{I}_{\{\Omega_2\}}$$

which is needed in the preceding estimates. Even if the heat conductivity in the conductor λ_1 is large, it cannot influence the estimates in corollaries 3.6 and 3.7. This is in contrast to the expected behaviour of the solutions of (3.14). That is why we propose an alternative model for the estimates of the stationary problem (3.15). It consists in reducing the domain Ω_{cr} to Ω_1 under an appropriate transformation of the boundary condition on Γ_β to $\partial \Omega_1$.

3.2.3. Transformation of the monotone boundary condition

To restrict the stationary problem (3.15) to Ω_1, we use relations from the case of rotational symmetry. Obviously, the transformed problem will not be equivalent to (3.15) if the domains Ω_1, Ω_2 are not circular. Nevertheless, it is a plausible idealization, since electric mains are rotationally symmetric in many cases. An a priori error analysis for approximation by rotationally symmetric geometries is outstanding.

The case of rotational symmetry

In addition to the previous notation we introduce

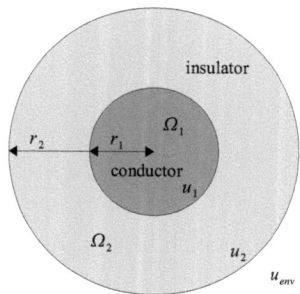

r_1 radius of the conductor

r_2 radius of the main

u_1 temperature at $\Gamma_1 = \partial B_{r_1}$

u_2 temperature at $\Gamma_2 = \partial B_{r_2}$.

Consider now the following cross-sectional boundary value problem

$$-\lambda_1 \Delta u = \frac{\rho_0 \left(1 + \alpha_\rho u_1\right) I^2}{|\Omega_1|^2} \quad \text{in } \Omega_1 = B_{r_1} \tag{3.16}$$

$$-\lambda_2 \Delta u = 0 \quad \text{in } \Omega_2 = B_{r_2} \setminus B_{r_1}$$

$$-\lambda_2 \frac{\partial u}{\partial n} = \alpha(u_2)\left(u_2 - u_{env}\right) \quad \text{on } \Gamma_2$$

Remark

For simplicity we assume that the resistivity $\rho = \rho_0 \left(1 - \alpha_\rho u\right)$ in (3.17) depends on the constant boundary temperature u_1 only. It is plausible as the temperature profiles in conductors are nearly constant. We refer to section 2.2 for the respective error estimate.

We use the rotationally symmetric form of the Laplace-operator in $\mathbb{R}^2$ to solve (3.16) and obtain $u_1 - u_2 = \frac{\rho_0 \left(1 + \alpha_\rho u_1\right) I^2}{2 \pi \lambda_2 |\Omega_1|} \ln\left(\frac{r_2}{r_1}\right)$. Moreover an integration of the boundary condition in (3.16) over Γ_2 yields $\int_{\Gamma_2} \lambda_2 \frac{\partial u}{\partial n} d\sigma = |\Gamma_2| \alpha(u_2)(u_2 - u_{env})$. A power comparison between the heat flux on the left hand side and the integrated source term $\int_{\Omega_1} \frac{\rho_0 \left(1 + \alpha_\rho u_1\right) I^2}{|\Omega_1|^2} dx$ in (3.16) gives

$$u_2 - u_{env} = \frac{\rho_0 \left(1 + \alpha_\rho u_1\right) I^2}{2 \pi r_2 \alpha(u_2) |\Omega_1|}.$$

Note that the Divergence Theorem cannot be applied here, since we have $u \notin C^1(\Omega_1 \cup \Omega_2)$. In fact it would yield the wrong result $u_2 - u_{env} = \frac{\lambda_2}{\lambda_1} \frac{\rho I^2}{2 \pi r_2 \alpha |\Omega_1|}$.

Definition of the ratio η

We want to replace the boundary condition $-\lambda_2 \frac{\partial u}{\partial n} = \alpha(u_2)(u_2 - u_{env})$ on Γ_2 in (3.16) by an energy conserving boundary condition on Γ_1. Therefore we introduce the ratio between the inner and outer boundary temperature $\eta := \frac{u_2 - u_{env}}{u_1 - u_{env}}$. Due to the previous formulas for $u_2 - u_1$ and $u_1 - u_{env}$ we have $\eta = \tilde{\eta}(u_2) = \frac{1}{1 + \alpha(u_2) r_2 \frac{\ln(r_2/r_1)}{\lambda_2}}$. This ratio depends on the outer boundary temperature u_2 which is adverse for a formulation of a boundary condition on Γ_1. Hence we consider the bijective map $t_{21} : (u_{env}, \infty) \to (u_{env}, \infty)$; $u_1 \mapsto u_2 = t_{21}(u_1)$. It is defined as the solution map of the equation

$$0 = \tilde{\eta}(u_2)(u_1 - u_{env}) - (u_2 - u_{env}) \qquad (3.17)$$

for given u_1, u_{env}. It maps the inner boundary temperature u_1 uniquely on the outer boundary temperature u_2 since we have

Lemma 3.1

Every $u_1 \in (u_{env}, \infty)$ admits a unique solution $u_2 \in (u_{env}, \infty)$ of (3.17).

Proof

Defining $F(u_2) := \tilde{\eta}(u_2)(u_1 - u_{env}) - (u_2 - u_{env})$ we have $F'(u_2) < -1$ for every $u_1, u_2 \in (u_{env}, \infty)$, which implies the assertion of Lemma 3.1. □

The following figure depicts the behaviour of F for $u_{env} = 50$, $u_1 = 100$, $\epsilon = 0.93$, $r_1 = 0.001$, $r_2 = 0.002$, $\lambda_2 = 0.2$.

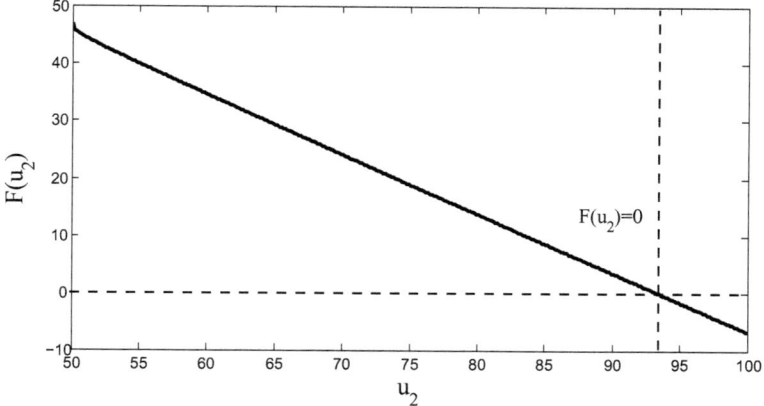

Equation (3.17) can be solved -i.e. t_{21} can be evaluated- via Newton's method.

Thus we finally define the ratio

$$\eta = \eta(u_1) = \tilde{\eta} \circ t_{21}(u_1) = \frac{1}{1 + \alpha(t_{21}(u_1))\, r_2 \frac{\ln(r_2/r_1)}{\lambda_2}}. \tag{3.18}$$

Energy conservating transformation

Due to the stationary process in (3.16) we have an equality between the heat flows on Γ_1 and Γ_2, i.e. $\int_{\Gamma_1} \lambda_1 \frac{\partial u}{\partial n}\, d\sigma = \int_{\Gamma_2} \lambda_2 \frac{\partial u}{\partial n}\, d\sigma$. Thus there holds $|\Gamma_1| \lambda_1 \frac{\partial u}{\partial n}|_{\Gamma_1} = |\Gamma_2| \lambda_2 \frac{\partial u}{\partial n}|_{\Gamma_2} \stackrel{(3.16)}{=} -|\Gamma_2|\, \alpha(u_2)\,(u_2 - u_{env})$ which implies $-\lambda_1 \frac{\partial u_1}{\partial n} = \frac{|\Gamma_1|}{|\Gamma_2|}\, \alpha(u_2)\,(u_2 - u_{env})$ on Γ_1. Here we observe an inconsistency between the presence of u_2 on the right hand side of the monotone boundary condition and its localization on Γ_1. Hence we apply the ratio η from (3.18) which gives

$$-\lambda_1 \frac{\partial u_1}{\partial n} = \frac{|\Gamma_2|}{|\Gamma_1|}\, \eta(u_1)\, \alpha(t_{21}(u_1))\,(u_1 - u_{env}). \tag{3.19}$$

Now we obtain an equivalent formulation of (3.16) restricted to the conductor domain Ω_1.

$$-\lambda_1 \Delta u = \frac{c_0\,(1 + \alpha_\rho u_1)\, I^2}{|\Omega_1|^2} \quad \text{in } \Omega_1$$

$$-\lambda_1 \frac{\partial u}{\partial n} = \frac{|\Gamma_2|}{|\Gamma_1|}\, \eta(u_1)\,(\alpha \circ t_{21})(u_1)\,(u_1 - u_{env}) \quad \text{on } \Gamma_1$$

Justified by the arguments at the beginning of this section, we apply the transformation (3.19) to the boundary value problem in (3.15)

$$-\lambda_1 \Delta u_{st} = \frac{\rho_0\,(1 + \alpha_\rho u_{st})\, I^2}{|\Omega_1|^2} \quad \text{in } \Omega_1 \times (-l_0, l_0) \tag{3.20}$$

$$-\lambda_1 \frac{\partial u_{st}}{\partial n} = \frac{|\partial \Omega_{cr}|}{|\partial \Omega_1|}\, \eta(u_{st})\,(\alpha \circ t_{21})(u_{st})\,(u_{st} - u_{env}) \quad \text{on } \partial \Omega_1 \times (-l_0, l_0)$$

$$\lambda_1 \frac{\partial u_{st}}{\partial n} = g_{N_1,1} \text{ on } \Gamma_{N_1} \cap \Omega_1\ ;\quad \lambda_1 \frac{\partial u_{st}}{\partial n} = g_{N_2,1} \text{ on } \Gamma_{N_2} \cap \Omega_1.$$

We abbreviate $\Omega_1 \times (-l_0, l_0) = \Omega_{1,l_0}$, $\partial\Omega_1 \times (-l_0, l_0) = \Gamma_\beta$; $\Gamma_{N_i} \cap \Omega_1 = \Gamma_{N_i}$, $g_{N_i,1} = g_{N_i}$, $i = 1, 2$ and $\Gamma_{N_1} \cup \Gamma_{N_2} = \Gamma_g$ concerning the transformed problem in the following.

3.2.4. Subresonance for the transformed problem (3.20) and its sensitivity and asymptotics for $\alpha_\rho \to 0$, $l \to \infty$

First we re-identify the general setting for (3.20) by

$$\Lambda = \lambda_1 \begin{pmatrix} 1 & 0 & 0 \\ 0 & 1 & 0 \\ 0 & 0 & 1 \end{pmatrix} \quad \text{i.e.} \quad \lambda_{min} = \lambda_1$$

$$\beta(u) = \frac{|\partial\Omega_{cr}|}{|\partial\Omega_1|} \eta(u) \, (\alpha \circ t_{21})(u) \, (u - u_{env}) \,, \quad g = g_{N_1} \mathbb{I}_{\Gamma_{N_1}} + g_{N_2} \mathbb{I}_{\Gamma_{N_2}} \,.$$

$$\text{and} \quad f = \frac{\rho_0 \, I^2}{|\Omega_1|^2} \,, \quad r = r(u) = \frac{\rho_0 \, I^2}{|\Omega_1|^2} u \quad \text{i.e.} \quad L_r = \frac{\rho_0 \, I^2}{|\Omega_1|^2}$$

The identification of c_β needs a detailed treatment.

Estimate for c_β

Assume that the monotone boundary transfer map $\beta(u) = \frac{|\partial\Omega_{cr}|}{|\partial\Omega_1|} \eta(u) \, (\tilde{\alpha} \circ t_{21})(u) \, (u - u_{env})$ is differentiable for $u \in [u_l, u_h]$ and $u_{env} \leq u_l$; where $\tilde{\alpha}$ denotes the truncation from (3.3). Then we have via the product rule

$$\beta'(s) = \frac{|\partial\Omega_{cr}|}{|\partial\Omega_1|} \left((\eta(s) \, (\tilde{\alpha} \circ t_{21})(s))' \, (s - u_{env}) + \eta(s) \, (\tilde{\alpha} \circ t_{21})(s) \right) .$$

Now observe that by the definition of η in (3.18) the map $s \mapsto \eta(s) \, (\alpha \circ t_{21})(s) =: G(s)$ is monotonically increasing for $s \in [u_l, u_h]$. For $u_{env} = 50$, $\epsilon = 0.93$, $r_1 = 0.001$, $r_2 = 0.002$, $\lambda_2 = 0.2$, the monotonicity of G is depicted in the following graph.-

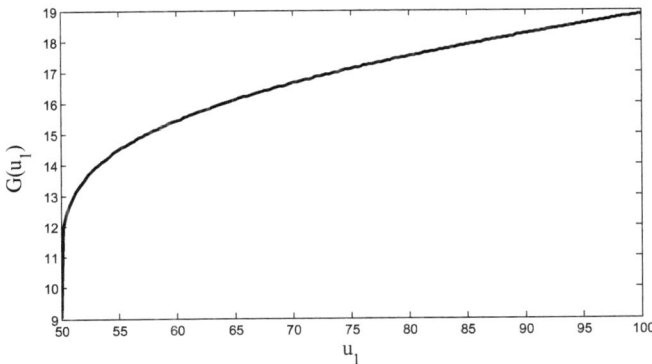

Thus we get

$$\inf_{s\in[u_l,u_h]} \beta'(s) \geq \frac{|\partial\Omega_{cr}|}{|\partial\Omega_1|}\eta(u_l)\,\alpha_l = \frac{|\partial\Omega_{cr}|}{|\partial\Omega_1|} \frac{\alpha_l}{1+r_2\,\alpha_l\,\frac{\ln(r_2/r_1)}{\lambda_2}}.$$

Hence we identify c_β with this lower bound; i.e.

$$c_\beta = \frac{|\partial\Omega_{cr}|}{|\partial\Omega_1|} \frac{\alpha_l}{1+r_2\,\alpha_l\,\frac{\ln(r_2/r_1)}{\lambda_2}} \qquad (3.21)$$

where $r_2 = diam(\Omega_{cr})/2$ and $r_1 = diam(\Omega_1)/2$.

Remark
Let $c_\beta^{in} = \frac{|\partial\Omega_{cr}|}{|\partial\Omega_1|} \frac{\alpha_l}{1+r_2\,\alpha_l\,\frac{\ln(r_2/r_1)}{\lambda_2}}$ denote the monotonicity constant of the transformed inner boundary map β on Γ_β and c_β^{out} the monotonicity constant of the original boundary map β on $\partial\Omega_{cr}\times(-l_0,l_0) = \Gamma_\beta$ in (3.15); here for a stationary identification of Λ, hence no heat capacity appears. For any realistic setting - e.g. $r_1 = 0.001$, $r_2 = 0.002$, $\lambda_2 = 0.2$ - we observe $c_\beta^{in} \approx \frac{|\partial\Omega_{cr}|}{|\partial\Omega_1|} c_\beta^{out}$. This is a special case of the damping property which reads in general as $c_\beta^{in} \geq c_\beta^{out}$. It means that a change of the boundary temperature changes the inner normal derivative more than the outer normal derivative. We describe it in detail in chapter 4.

Now we formulate sufficient conditions for subresonance for the transformed problem.

Corollary 3.8
Let $\alpha_\rho < \frac{\lambda_1 |\Omega_1|^2}{\rho_0 I^2 c_\star^2}$. Then there exists a unique stationary solution $u_{st} \in H^1(\Omega_{l_0})$ of (3.20) which is bounded by

$$\left(\lambda_1 - \frac{\rho_0 I^2 |\alpha_\rho|}{|\Omega_1|^2} c_\star^2\right) \|u_{st}\|_{\star,l_0} \leq C_{\rho,g} + C_\alpha$$

where $C_{\rho,g} = c_\star \sqrt{2 l_0} \frac{\rho_0 I^2}{|\Omega_1|^{3/2}} + C \|g\|_{L^2(\Gamma_g)}$ and $C_\alpha = \sqrt{\lambda_1 |\Gamma_\beta| \alpha_l} |u_{env}|$.
$C := \|\tau\|_{tr} = \sup\limits_{\|v\|_{\star,l_0} \leq 1} \|\tau(v)\|_{L^2(\Gamma_g)}$ denotes the norm of the trace map $\tau : H^1(\Omega_{1,l_0}) \to L^2(\Gamma_g)$. and $c_\star$ the Friedrichs constant for Ω_{1,l_0}.

If we consider the cross-sectional problem

$$-\lambda_1 \Delta u = \frac{\rho_0 (1 + \alpha_\rho u) I^2}{|\Omega_1|^2} \quad \text{in } \Omega_1 \qquad (3.22)$$

$$-\lambda_1 \frac{\partial u}{\partial n} = \frac{|\partial \Omega_{cr}|}{|\partial \Omega_1|} \eta(u) (\alpha \circ t_{21})(u) (u - u_{env}) \quad \text{on } \partial \Omega_1$$

the Friedrichs-constant reads as

$$c_{\star,1} = \sqrt{diam(\Omega_1) \lambda_1 \frac{|\partial \Omega_1|}{|\partial \Omega_{cr}|} \left(\frac{1}{\alpha_l} + r_2 \frac{\ln(r_2/r_1)}{\lambda_2}\right)} \qquad (3.23)$$

for the small scale case $diam(\Omega_1) \leq \frac{\lambda_1}{c_\beta} = \lambda_1 \frac{|\partial \Omega_1|}{|\partial \Omega_{cr}|} \left(\frac{1}{\alpha_l} + r_2 \frac{\ln(r_2/r_1)}{\lambda_2}\right)$.
With that Corollary 3.8 simplifies to

Let $\alpha_\rho < \frac{\lambda_1 |\Omega_1|^2}{\rho_0 I^2 c_\star^2}$. Then there exists a unique stationary solution $u \in H^1(\Omega_1)$ of (3.22) which is bounded by

$$\left(\lambda_1 - \frac{\rho_0 I^2 |\alpha_\rho|}{|\Omega_1|^2} c_{\star,1}^2\right) \|u\|_\star \leq C_\rho + C_\alpha$$

where $C_\rho = c_{\star,1} \frac{\rho_0 I^2}{|\Omega_1|^{3/2}}$ and $C_\alpha = \sqrt{\lambda_1 |\partial\Omega_{cr}| \alpha_l} |u_{env}|$.

Remark
In applications we have $\lambda_1 \gg \lambda_2$. Observe that this yields a distinct extension of the subresonance condition and a much smaller bound on $\|u_{st}\|_{\star,l_0}$; i.e. an improvement of Corollary 3.6. This improvement continues in the sensitivity estimate for $\alpha_\rho \to 0$ and the asymptotic estimate for $l \to \infty$.

Sensitivity for $\alpha_\rho \to 0$ and asymptotics for $l \to \infty$

We consider the boundary value problem (3.20) for $\alpha_\rho = 0$, i.e.

$$-\lambda_1 \Delta u = \frac{\rho_0 I^2}{|\Omega_1|^2} \quad \text{in } \Omega_{1,l_0} \tag{3.24}$$

$$-\lambda_1 \frac{\partial u}{\partial n} = \frac{|\partial\Omega_{cr}|}{|\partial\Omega_1|} \eta(u) (\alpha \circ t_{21})(u) (u - u_{env}) \quad \text{on } \Gamma_\beta$$

$$\lambda_1 \frac{\partial u}{\partial n} = g \quad \text{on } \Gamma_g.$$

The solution exists due to Corollary 3.8. The asymptotics of solutions u_{st} of (3.20) for $\alpha_\rho \to 0$ to solutions u of (3.24) is given by

Corollary 3.9
Assume $\alpha_\rho < \frac{\lambda_1 |\Omega_1|^2}{\rho_0 I^2 c_\star^2}$. Then the following estimate holds

$$\|u_{st} - u\|_{\star,l} \leq \frac{|\alpha_\rho| \rho_0 I^2 c_\star}{\lambda_1 |\Omega_1|^2 - |\alpha_\rho| \rho_0 I^2 c_\star^2} \|u\|_{L^2(\Omega_{1,l_0})}.$$

Remark
The difference between Corollary 3.4 and Corollary 3.9 is determined by different monotonicity constants c_β for the monotone boundary mapping β and thus by different Friedrichs-constants $c_\star$. Here we observe that the larger c_β and thus smaller $c_\star$ in Corollary 3.9 even improves the estimate in Corollary 3.4. I.e. for a realistic choice of material parameters, the insulator reduces the influence the temperature coefficient α_ρ and extends the subresonant state.

We will treat this effect quantitatively in section 3.2.5.

Finally we consider the cross-sectional problem

$$-\lambda_1 \Delta \bar{u} = \frac{\rho_0 I^2}{|\Omega_1|^2} \quad \text{in } \Omega_1 \qquad (3.25)$$

$$-\lambda_1 \frac{\partial \bar{u}}{\partial n} = \frac{|\partial \Omega_{cr}|}{|\partial \Omega_1|} \eta(\bar{u})\,(\alpha \circ t_{21})(\bar{u})\,(\bar{u} - u_{env}) \quad \text{on } \partial \Omega_1$$

whose solution exists uniquely due to the cross-sectinal variant of Corollary 3.8. The extension of the cross-sectional data to $\Omega_l \subset \mathbb{R}^3$ is given by

$$f_\infty = \bar{f} = \frac{\rho_0 I^2}{|\Omega_1|^2}, \quad \Lambda_\infty = \lambda_1 \begin{pmatrix} 1 & 0 & 0 \\ 0 & 1 & 0 \\ 0 & 0 & 1 \end{pmatrix}, \quad \text{i.e. } \lambda_{min} = \lambda_{ddmax} = \lambda_1$$

and $u_\infty(x_1, x_2, x_3) = \bar{u}(x_1, x_2)$.

The related cylinder boundary value problem reads as (3.20) w.r.t. Ω_l. We show the convergence of solutions of (3.20) - labeled $(u_l)_{l>0}$ - towards the extended solution u_∞ of the cross-sectional problem (3.25) for large l.

Corollary 3.10

Let u_l denote the solution of (3.20) and u_∞ the extended solution of (3.25). Then, for $l_0 < l$ there holds

$$\lambda_1 \|u_l - u_\infty\|_{*,l_0} \leq C \exp\left(\frac{-(l-l_0)}{c_{*,1}}\right) \|g\|_{L^2(\Gamma_g)}.$$

$C := \|\tau\|_{tr} = \sup_{\|v\|_{*,l_0} \leq 1} \|\tau(v)\|_{L^2(\Gamma_g)}$ denotes the norm of the trace map $\tau : H^1(\Omega_1, l_0) \to L^2(\Gamma_g)$. and $c_{*,1}$ denotes the generalized Friedrichs-constant for Ω_1 w.r.t. the transformed c_β. It is given by (3.23) for the small scale case. Note that Theorem 2.3 applied to the non-transformed problem (3.15) would give a worse estimate in Corollary 3.10; namely with $\tilde{c}_\lambda = \frac{\lambda_{ddmax}}{\lambda_{min}} c_{*,cr} = \frac{\lambda_1}{\lambda_2} c_{*,cr} \gg c_{*,1} = c_\lambda$, since $\lambda_1 \gg \lambda_2$.

3.2.5. Remarks on the transformation of problem (3.15)

Energy conservation and time independence

The transformation proposed in section 3.2.3 applies to the stationary problem (3.15) only. The reason is, that the energy conservation argument

$$\int_{\Gamma_1} \lambda_1 \frac{\partial u}{\partial n}\, d\sigma = \int_{\Gamma_2} \lambda_2 \frac{\partial u}{\partial n}\, d\sigma$$

is a stationary one. To obtain a time dependent energy conservation argument, we have to regard the capacitary absorption of heat in the insulator, i.e.

$$\int_{\Gamma_1} \lambda_1 \frac{\partial u}{\partial n}\, d\sigma = \int_{\Gamma_2} \lambda_2 \frac{\partial u}{\partial n}\, d\sigma + \int_{\Omega_2} \gamma_2 \frac{\partial u}{\partial t}\, dx .$$

This can be used for a transformation of the time dependent problem (3.14) and thus for an a priori improvement the asymptotics for $t \to \infty$ of Corollary 3.7 which is outstanding.

Influence of the insulation on the change of the subresonant state

Comparing Corollaries 3.4 and 3.5 with Corollaries 3.9 and 3.10 we observe that the insulation changes the generalized Friedrichs constant only. A decrease of $c_\star$ extends the subresonant state and improves the associated estimates for $\alpha_\rho \to 0$ and $l \to \infty$. An increase of $c_\star$ effects the contrary. The change of $c_\star$ is caused by a change of the monotonicity constant c_β. The following Proposition considers the uninsulated cross-sectional problem

$$-\lambda_1 \Delta u = \frac{\rho_0 (1 + \alpha_\rho u) I^2}{|\Omega_1|^2} \quad \text{in } \Omega_1 \tag{3.26}$$

$$-\lambda_1 \frac{\partial u}{\partial n} = \alpha(u)(u - u_{env}) \quad \text{on } \partial\Omega_1$$

and its insulated and transformed counterpart in (3.22). As before we use $\Gamma_1 = \partial\Omega_1$, $\Gamma_2 = \partial\Omega_2 \setminus \delta\Omega_1 = \partial\Omega_{cr}$, $r_1 = diam(\Omega_1)/2$, $r_2 = diam(\Omega_2)/2$ for the perimeters and diameters of the conductor and the insulator domain.

Proposition 3.1

Assume that the insulation parameters in (3.22) fulfill the relation

$$\frac{\lambda_2}{\alpha_l} \geq \frac{r_2 \ln(r_2/r_1)}{\frac{|\Gamma_2|}{|\Gamma_1|} - 1}.$$

Then the insulation extends the subresonant state of (3.26). The complementary relation $\frac{\lambda_2}{\alpha_l} < \frac{r_2 \ln(r_2/r_1)}{\frac{|\Gamma_2|}{|\Gamma_1|}-1}$ causes a contraction of the s.r.s. in (3.26).

Proof

The assertion follows immediately from the comparison of $c_\beta = \alpha_l$ in (3.4) (stationary interpretation) and $c_\beta = \frac{|\Gamma_2|}{|\Gamma_1|} \frac{\alpha_l}{1 + r_2 \alpha_l \frac{\ln(r_2/r_1)}{\lambda_2}}$ in (3.21) for the insulated and transformed case.

A distinction between the large scale and the small scale case is not necessary, since $c_\star$ does not explicitly depend on c_β in the large scale case. □

Proposition 3.1 gives an orientation whether an insulation improves the thermal behavior - i.e. extends the subresonant state - of an electric cable or not; depending on the geometrical and physical properties of the insulation.

Nevertheless, Proposition 3.1 compares the rather restricitve subresonance conditions of Corollaries 3.4 and 3.9 only. Now we compare the temperature u_1 of the pure conductor problem and the inner temperature $\tilde{u}_1$ of the insulator-conductor problem directly. Hereto we suppose that the generated heat power P of the pure problem and the heat power $\tilde{P}$ of the insulated problem are equal. This implies

$$\alpha(u_1)(u_1 - u_{env})|\Gamma_1| = \tilde{\alpha}(\tilde{u}_2)(\tilde{u}_2 - u_{env})|\Gamma_2| = \tilde{\alpha}(\tilde{u}_2)\eta(\tilde{u}_1 - u_{env})|\Gamma_2|.$$

Here, α, $\tilde{\alpha}$ denote the heat transfer coefficient on the conductor boundary Γ_1 and the heat transfer coefficient on the outer insulator boundary Γ_2.

Consider now the ratio $\psi = \frac{\tilde{u}_1 - u_{env}}{u_1 - u_{env}} = \frac{\alpha(u_1)|\Gamma_1|}{\tilde{\alpha}(\tilde{u}_2)|\Gamma_2|\eta}$. Then, plausibly, $\psi \leq 1$ describes a cooling effect and $\psi > 1$ a heating effect of the insulation. Hence, using the definition of η, we obtain

$$\tilde{\alpha}(\tilde{u}_2)\,|\Gamma_2| \geq |\Gamma_1|\,\alpha(u_1)\left(1+\tilde{\alpha}(\tilde{u}_2)\,r_2\frac{\ln(r_2/r_1)}{\lambda_2}\right) \qquad (3.27)$$

$$\tilde{\alpha}(\tilde{u}_2)\,|\Gamma_2| < |\Gamma_1|\,\alpha(u_1)\left(1+\tilde{\alpha}(\tilde{u}_2)\,r_2\frac{\ln(r_2/r_1)}{\lambda_2}\right) \qquad (3.28)$$

The cooling condition (3.27) and the heating condition (3.28) are implicit and have to be evaluated in specific cases for a known range of temperatures u_1, $\tilde{u}_2$. Observe that Proposition 3.1 yields a sufficient condition $\frac{\lambda_2}{\alpha_l} < \frac{r_2\ln(r_2/r_1)}{\frac{|\Gamma_2|}{|\Gamma_1|}-1}$ for the heating effect in (3.28).

Concluding Remarks

The transformation in section 3.2.3 makes use of an approximately rotationally symmetric shape of cross sections of electric cables to replace the outer data of the insulator. In section 4.2.2 we consider the cross-sectional problem

$$\begin{aligned}
-\lambda_1\,\Delta u &= \frac{\rho_0\,(1+\alpha_\rho u)\,I^2}{|\Omega_1|^2} \quad \text{in } \Omega_1 \\
-\lambda_2\,\Delta u &= 0 \quad \text{in } \Omega_2 \\
-\lambda_2\,\frac{\partial u}{\partial n} &= \alpha(u)\,(u-u_{env}) \quad \text{on } \partial\Omega_2
\end{aligned}$$

on the insulator domain Ω_2 only. Hereto we will replace the source term $\frac{\rho_0\,(1+\alpha_\rho u)\,I^2}{|\Omega_1|^2}$ in Ω_1 by a heat flow over $\partial\Omega_1$ using approximation by rotational symmetry.

4. Treatment by nonlinear boundary integral equations

In this chapter we consider the heat transfer in electric cables on the boundaries of the respective domains and use the notation of chapter 3. As a basis we investigate a cross-sectional stationary problem; i.e. the reduction of the full problem (3.1) or (3.14) via $t \to \infty$ and $l \to \infty$ treated in chapter 3. On the other hand we do not neglect the temperature dependence of the resistivity ρ completely. We rather restrict it to the conductor boundary as described in section 3.1.4.

In section 4.1 we deploy an equivalent boundary integral equation for the cross-sectional problem using single and double layer potential operators. We use the Theorem of Browder and Minty on monotone operators to prove existence and uniqueness of the solution of the boundary integral equation (b.i.e.). Then we transform the nonlinear b.i.e. to a fixed point equation on an appropriate Sobolev space and compute the solution via an iterative method presented for abstract Hilbert spaces by Browder and Petryshyn in [13] and by Brézis and Sibony in [11]. We illustrate this method for rotationally symmetric conductors where the boundrary temperature reduces to a constant value.

In section 4.2 we consider an insulated domain and formulate the heat transfer problem on the insulator domain only. Here the maximum principle for harmonic functions implies that the temperatures at the boundary of the insulator domain are the extremal and thus relevant unknowns. This gives rise to treat the problem by boundary integral equations on multiply connected insulator domains. For this purpose we extend the analysis of section 4.1 to matrix valued boundary integral operators. Here, as in the simply connected case, the strong monotonicity of the Poincaré-Steklov operator of the underlying boundary value problem is essential. In this context we introduce an abstract property for boundaries of multiply connected domains - the damping property. This property enables us to verify the strong monotonicity of the Poincaré-Steklov operator independently from the conductor parameters, i.e. just using the outer boundary condition.

Finally we deal with the case of rotational symmetry. Here the boundary integral operators reduce to matrices which can be computed explicitly. Thus we obtain the solution as the limit of an iterative sequence of vectors.

We emphasize that the presentation of the specific example of heat transfer in electric cables does not obstruct an application of the boundary integral approach to other problems governed by elliptic equations.

4.1. Boundary integral approach for uninsulated cables

4.1.1. Setup of the problem

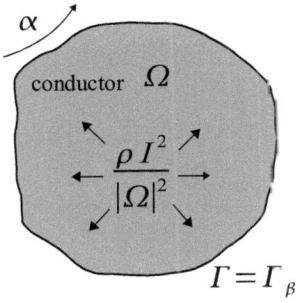

Let $\Omega \subset \mathbb{R}^2$ have a Lipschitz boundary $\Gamma = \partial \Omega$. We consider the following cross-sectional boundary value problem

$$-\lambda \Delta u_{st} = \frac{\rho I^2}{|\Omega|^2} \quad \text{in } \Omega \qquad (4.1)$$

$$-\lambda \frac{\partial u_{st}}{\partial n} = \alpha(u_{st})(u_{st} - u_{env}) \quad \text{on } \Gamma.$$

Using the model of a linear temperature dependent resistivity from section

3.1.1 we have $\rho(u) = \rho_0 (1 + \alpha_\rho u)$. By comparatively large heat conductivity λ, small differences in temperature in the conductor material can be expected. This motivates a restriction of the dependence to a mean value boundary temperature.

Restriction of the temperature dependence of ρ to $\bar{u}$

Following section 3.1.4 we approximate (4.1) by a Poisson-Equation

$$-\lambda \Delta u = \frac{\rho_0 I^2 (1 + \alpha_\rho \bar{u})}{|\Omega|^2} \quad \text{in } \Omega \tag{4.2}$$

$$-\lambda \frac{\partial u}{\partial n} = \alpha(u) (u - u_{env}) \quad \text{on } \Gamma.$$

with a suitably chosen mean value temperature $\bar{u} \in \mathbb{R}$. The existence and uniqueness result for (4.1) combined with an error estimate for the approximation by (4.2) reads as

Assume $\alpha_\rho < \frac{\lambda |\Omega|^2}{\rho_0 I^2 c_\star^2}$. Then there exists a unique solution $u_{st} \in H^1(\Omega)$ of (4.1) which is approximated by the solution of (4.2) via

$$\|u_{st} - u\|_\star \leq C_{\alpha_\rho} \|u - \bar{u}\|_{L^2(\Omega)} \quad \text{where} \quad C_{\alpha_\rho} = \frac{|\alpha_\rho| \rho_0 I^2 c_\star}{\lambda |\Omega|^2 - |\alpha_\rho| \rho_0 I^2 c_\star^2}.$$

To obtain $\bar{u}$ for the Poisson datum in (4.2) which is a priori known we cannot take the error minimizing mean value $\frac{1}{|\Omega|} \int_\Omega u \, dx$ as proposed in section 2.2.2. We rather use the implicitly defined energy conservating mean value which is constant in space and solves the algebraic equation (3.8), i.e. the solution of

$$\frac{\rho_0 I^2}{|\Omega|} (1 + \alpha_\rho \bar{u}) = |\partial \Omega| \alpha(\bar{u}) (\bar{u} - u_{env}). \tag{4.3}$$

It exists uniquely for $\alpha_\rho < \frac{|\partial \Omega| |\Omega| \alpha_l}{\rho_0 I^2}$ which is implied by the subresonance condition $\alpha_\rho < \frac{\lambda |\Omega|^2}{\rho_0 I^2 c_\star^2}$. The solution $\bar{u}$ can be found e.g. by Newton's method or via a fixed point iteration applied to the equation

$$\bar{u} = u_{env} + \frac{\rho_0 I^2}{|\Omega| |\partial\Omega| \alpha(\bar{u})} (1 + \alpha_\rho \bar{u}) =: \zeta_m(\bar{u}). \qquad (4.4)$$

Proposition 4.1 (Convergence of the fixed point iteration)
Let $\alpha : [u_{env}, \infty) \to [\alpha_l, \alpha_h]$; denote the truncated heat transfer coefficient from (3.3) and let the truncation yield the relation

$$\left| \frac{1 + \alpha_\rho u_2}{\alpha(u_2)} - \frac{1 - \alpha_\rho u_1}{\alpha(u_1)} \right| \le \frac{\alpha_\rho}{\alpha_l} |u_2 - u_1| \; ; \; u_1, u_2 \in [u_{env}, \infty).$$

Moreover let the relation $\alpha_\rho < \frac{|\partial\Omega| |\Omega| \alpha_l}{\rho_0 I^2}$ hold. Define the iterative sequence $(\bar{u}^{(n)})_{n \in \mathbb{N}} \subset \mathbb{R}$ by $\bar{u}^{(n+1)} := \zeta_m(\bar{u}^{(n)})$.
Then, for any initial value $\bar{u}^{(1)} \in [u_{env}, \infty)$ the iterative sequence $(\bar{u}^{(n)})_{n \in \mathbb{N}}$ converges to the unique solution $\bar{u}$ of (4.3) with the following rate of convergence

$$|\bar{u}^{(n)} - \bar{u}| \le \frac{q^n}{1-q} |\bar{u}^{(2)} - \bar{u}^{(1)}| \quad \text{where } q := \frac{\rho_0 \alpha_\rho I^2}{\alpha_l |\Omega| |\partial\Omega|}.$$

Proof
We show that the assumption $\rho_0 \alpha_\rho I^2 < |\partial\Omega| |\Omega| \alpha_l$ yields global contractivity of the map $\zeta_m : [u_{env}, \infty) \to [u_{env}, \infty)$. We have

$$|\zeta(s_2) - \zeta(s_1)| = \frac{\rho_0 I^2}{|\Omega| |\partial\Omega|} \left| \frac{1 + \alpha_\rho s_2}{\alpha(s_2)} - \frac{1 + \alpha_\rho s_1}{\alpha(s_1)} \right| \le \underbrace{\frac{\rho_0 \alpha_\rho I^2}{\alpha_l |\Omega| |\partial\Omega|}}_{=q} |s_2 - s_1|.$$

Thus existence and uniqueness of a solution of (4.3) and convergence of the iterative sequence $(\bar{u}_n)_{n \in \mathbb{N}}$ follow by Banach's fixed point theorem, (A.1). □

Remarks
(i) If the truncation is suitably chosen, an example for a heat transfer coefficient which fulfills the requirements of Proposition 4.1 is given in section 3.1.5

(ii) In section 4.2.2 we will use another approach and replace the source term $\frac{\rho_0 (1+\alpha_\rho u) I^2}{|\Omega|}$ by - a not a priori known - temperature dependent heat flow over the boundary Γ.

4.1.2. Equivalent formulation by a nonlinear boundary integral equation

In the following we are concerned with the temperature on the boundary of the conductor domain. Using Green's representation formula we derive an equivalent boundary integral equation for $\Gamma = \partial\Omega$ that includes the nonlinear boundary condition in (4.2).
Starting from $-\Delta w = f$ in Ω the representation formula yields

$$w(\tilde{x}) = \int_\Omega F(\tilde{x} - y) f(y) \, dy \qquad (4.5)$$
$$+ \int_\Gamma \left(w(y) \frac{\partial}{\partial n_y} F(\tilde{x} - y) - \frac{\partial w(y)}{\partial n_y} F(\tilde{x} - y) \right) ds_y$$

for $\tilde{x} \in \Omega$ where $F(z) := \frac{1}{2\pi} \ln(|z|)$ denotes the fundamental solution of the Laplace-equation in $\mathbb{R}^2 \setminus \{0\}$.

To avoid a domain discretization of Ω in a possible numerical treatment, we transform the Newton potential $(\mathcal{N}f)(\tilde{x}) = \int_\Omega F(\tilde{x} - y) f(y) \, dy$, $\tilde{x} \in \Omega$ to a boundary integral operator.
Due to the restriction of ρ we have $f = \frac{\rho_0 I^2 (1+\alpha_\rho \bar{u})}{\lambda |\Omega|^2} = const.$ This allows an easy representation of $\mathcal{N}$ as a boundary integral via Gauß' Divergence Theorem and the fundamental solution for the biharmonic equation.

Lemma 4.1
A boundary integral formulation of $\mathcal{N}$ for constant densities f is given by $(\mathcal{N}f)(\tilde{x}) = -\int_\Gamma f \frac{\partial}{\partial n_y} F_b(\tilde{x} - y) \, ds_y$, $\tilde{x} \in \Omega$ where $F_b(z) := \frac{|z|^2}{8\pi} (\ln|z| - 1)$ denotes the fundamental solution of $\Delta^2 v = 0$ in $\mathbb{R}^2 \setminus \{0\}$.

Proof

F_b fulfills the relation $\Delta F_b = F$ in $\mathbb{R}^2 \setminus \{0\}$ where F is the fundamental solution of the Laplace equation. Thus we have

$$(\mathcal{N}f)(\tilde{x}) = \int_\Omega f \,\Delta F_b(\tilde{x}-y)\,dy = -\int_\Gamma f \frac{\partial}{\partial n_y} F_b(\tilde{x}-y)\,ds_y$$

by Gauß' Divergence Theorem. $\square$

The representation of $\mathcal{N}$ by boundary integrals applied to non-constant Poisson data f can be found in [65].

Jump relations and mapping properties for the boundary integral operators

Now the representation formula in (4.5) gives

$$w(\tilde{x}) = \int_\Gamma f \frac{\partial}{\partial n_y} F_b(\tilde{x}-y)\,ds_y + \int_\Gamma \left(w(y) \frac{\partial}{\partial n_y} F(\tilde{x}-y) - \frac{\partial u(y)}{\partial n_y} F(\tilde{x}-y) \right) ds_y.$$

Assume $\Gamma \in C^2$ and consider the limit $\Omega \ni \tilde{x} \to x \in \Gamma$. Then the jump relations of potential theory ([42], Sec. II) yield the boundary integral equation for $x \in \Gamma$. $\frac{u(x)}{2} = \int_\Gamma \left(f \frac{\partial}{\partial n_y} F_b(x-y) + u(y) \frac{\partial}{\partial n_y} F(x-y) - \varphi F(x-y) \right) ds_y.$

This equation reads as

$$0 = \mathcal{K}_b(f) + \left(\mathcal{K} - \frac{I}{2}\right)(u) + \mathcal{S}(\varphi) \tag{4.6}$$

where - if there is no risk of confusion - $u = w|_\Gamma$ denotes the Dirichlet data and $\varphi = \frac{\partial w}{\partial n}|_\Gamma$ the Neumann data of u.

Following singular boundary integral operator theory ([45], [48], [60]) we define the continuous mappings: the single layer operator $\mathcal{S} : H^{-1/2}(\Gamma) \to H^{1/2}(\Gamma)$, the double layer operator $\mathcal{K} : H^{1/2}(\Gamma) \to H^{1/2}(\Gamma)$ and the Bi-

Laplace double layer operator $\mathcal{K}_b : H^{-1/2}(\Gamma) \to H^{3/2}(\Gamma)$ by

$$(\mathcal{S}\varphi)(x) = -\int_\Gamma \varphi(y)\, F(x-y)\, \mathrm{d}s_y$$

$$(\mathcal{K}u)(x) = \int_\Gamma u(y)\, \frac{\partial}{\partial n_y} F(x-y)\, \mathrm{d}s_y$$

$$(\mathcal{K}_b f)(x) = -\int_\Gamma f\, \frac{\partial}{\partial n_y} F_b(x-y)\, \mathrm{d}s_y\,.$$

Here $\mathcal{K}_b$ results from the transformation of the Newton potential $\mathcal{N}$ to the boundary Γ. The jump relation for $\mathcal{K}_b$ is given in [65]. As the constant poisson datum is given by $f = \frac{\rho_0\,(1+\alpha_\rho\,\bar{u})\,I^2}{|\Omega|^2}$, we treat $\mathcal{K}_b f \in H^{3/2}(\Gamma)$ as a known function in the following.

Deployment of the nonlinear equation via the Hammerstein operator

Consider the map $h : \mathbb{R} \to \mathbb{R}$, $h(s) := \frac{\alpha(s)}{\lambda}(s - u_{env})$ from the boundary condition in (4.2). Analogously to section 2.1.1 we define the superposition operator $\Phi(u)(x) := h(u(x))$. Since we have $d = 2$, the continuity of h suficess to ensure the mapping property $\Phi : H^{1/2}(\Gamma) \to H^{-1/2}(\Gamma)$, [3].
Replacing the Neumann datum $-\varphi = \Phi(u)$ in (4.6) yields the Hammerstein operator $\mathcal{S} \circ \Phi : H^{1/2}(\Gamma) \to H^{1/2}(\Gamma)$ and the boundary integral equation

$$0 = \mathcal{K}_b(f) + \left(\mathcal{K} - \frac{I}{2} - \mathcal{S} \circ \Phi\right)(u) \quad \text{in} \quad H^{1/2}(\Gamma) \tag{4.7}$$

By previous considerations this equation is equivalent to the boundary value problem in (4.2).

Now there are two options to give an existence and uniqueness argument for solutions $u \in H^{1/2}(\Gamma)$ of (4.7). The first - indirect - one uses the subresonance Theorem 2.1 applied to (4.2). For a sufficiently smooth boundary, e.g. $\Gamma \in C^2$, there is a continuous trace mapping operator $\gamma : H^1(\Omega) \to H^{1/2}(\Gamma)$, which maps the solution of (4.2) to the wanted Dirichlet data in (4.7); see e.g. [4], prop. 5.6.3. Again we note that this method has the disadvantage of a domain discretization of Ω, if we want to treat (4.7) numerically.

4.1.3. Existence and Uniqueness of a solution of the nonlinear boundary integral equation

Therefore we give a direct existence and uniqueness argument for (4.7) in this section. It uses conditions for the Dirichlet-to-Neumann map Φ same with the condititons for the monotone boundary map β in Theorem 2.1. In addition we will need a scaling condition for Ω which ensures the invertibility of the single layer operator S. It is needed due to the specific structure of the fundamental solution $F(z) := \frac{1}{2\pi} \ln(|z|)$ which is the defining component of S. We formulate the following assumptions

(A1) *Scaling:* $diam(\Omega) < 1$
This assumption can be arranged without loss of generality and implies that $S : H^{-1/2}(\Gamma) \to H^{1/2}(\Gamma)$ is an isomorphism and $H^{-1/2}$-elliptic ([44]).

(A2) *Monotonicity and growth condition on h:*
We require that $h : \mathbb{R} \to \mathbb{R}$ is continuous such that the respective superposition operator $\Phi(u)(x) := h(u(x))$ maps $H^{1/2}(\Gamma)$ continuously to $H^{-1/2}(\Gamma)$. Moreover h shall satisfy the following monotonicity estimate

$$\exists\, c_{min} > 0 : \frac{h(s_1) - h(s_2)}{s_1 - s_2} \geq c_{min} \quad \text{for } s_1 \neq s_2$$

Such that $\Phi : H^{1/2}(\Gamma) \to H^{-1/2}(\Gamma)$ is strongly monotonous.

Remarks
(i) The mapping property for Φ in (A2) can be arranged without any growth condition on h since $\Omega \subset \mathbb{R}^2$, see section 2.1.1 and [3].
(ii) Observe that in section 2.1.1 h and c_{min} correspond to the monotone

boundary map $\frac{\beta}{\lambda}$ and the monotonicity constant $\frac{c_\beta}{\lambda}$ respectively.

Considering the homogeneous equation $\Delta w_0 = 0$ in Ω, we note that (4.6) reads as $\mathcal{S}(\varphi) = \left(\frac{I}{2} - \mathcal{K}\right)(u)$. Now condition (A1) allows to define the continuous Steklov-Poincaré operator $\mathcal{P} : H^{1/2}(\Gamma) \to H^{-1/2}(\Gamma)$ with

$$\mathcal{P} := \mathcal{S}^{-1} \circ \left(\frac{I}{2} - \mathcal{K}\right).$$

It maps the Dirichlet data $u \in H^{1/2}(\Gamma)$ of harmonic functions to the Neumann data $\varphi \in H^{-1/2}(\Gamma)$; see e.g. [60] for detailed considerations. Thus we apply $\mathcal{S}^{-1}$ to (4.7) and obtain

$$\mathcal{P}(u) + \Phi(u) = g \;,\; u \in H^{1/2}(\Gamma) \qquad (4.8)$$

where $g = \left(\mathcal{S}^{-1} \circ \mathcal{K}_b\right)(f) \in H^{-1/2}(\Gamma)$.

The main motivation for this Steklov-Poincaré representation is the separation of the nonlinearity Φ and the single-layer operator $\mathcal{S}$. Thus we are able to use the monotonicity assumption directly for the monotonicity of the left hand side of (4.8).

Theorem 4.1
Assume that (A1) and (A2) are satisfied. Then, for every $g \in H^{-1/2}(\Gamma)$ there exists a unique solution $u \in H^{1/2}(\Gamma)$ of (4.8) which is bounded by

$$\|u\|_{H^{1/2}(\Gamma)} \leq c_{emb}^2 \left(\|g\|_{H^{-1/2}(\Gamma)} + \sqrt{|\Gamma|}\, |h(0)|\right).$$

Remark
c_{emb} denotes the constant of the trace embedding between $H^{1/2}(\Gamma)$ and $H^1(\Omega)$; w.r.t. $\|\cdot\|_\star$; i.e. $\|\cdot\|_{H^{1/2}(\Gamma)} \leq c_{emb} \|\cdot\|_\star$.
Here $\|w\|_\star^2 = \|\nabla w\|_{L^2(\Omega)}^2 + c_{min} \|w\|_{L^2(\Gamma)}^2$ denotes the physically consistent norm on $H^1(\Omega)$. Analogously to section 2.1.1 it is equivalent to the canonical norm on $H^1(\Omega)$. Defining $\|u\|_{H^{1/2}(\Gamma)} := \inf\{\|v\|_\star : v|_\Gamma = u\}$, we can set $c_{emb} = 1$.

Proof of Theorem 4.1
(i) Existence and Uniqueness
Here we follow partly the proof of Theorem 2 in [72].
Consider $Au = g$ in $H^{-1/2}(\Gamma)$ with $A = \mathcal{P} + \Phi$. Due to the properties of $\mathcal{P}$ and Φ it is obvious that the operator $A : H^{1/2} \to H^{-1/2}$ is hemicontinuous. Thus it remains to show that A is strongly monotonous and the assertion follows by the Theorem of Browder and Minty for monotone operators. The linearity of the Steklov-Poincaré operator gives

$$\langle Au - Av, u - v \rangle = \langle \mathcal{P}(u-v), u-v \rangle + \langle \Phi(u) - \Phi(v), u-v \rangle.$$

Let $w_0 \in H^1(\Omega)$ denote the harmonic extension of the Cauchy-data $(u-v) \in H^{1/2}(\Gamma)$ and $\mathcal{P}(u-v) \in H^{-1/2}(\Gamma)$ to Ω. It is given uniquely by Green's representation formula. Then the Divergence Theorem implies

$$\langle \mathcal{P}(u-v), u-v \rangle = \int_\Gamma \frac{\partial(u-v)}{\partial n}(u-v)\,ds$$
$$= \int_\Omega \mathrm{div}\,(\nabla w_0\, w_0)\,dx = \int_\Omega |\nabla w_0|^2\,dx.$$

On the other hand the strong monotonicity of the superposition operator Φ yields $\langle \Phi(u) - \Phi(v), u-v \rangle \geq c_{min} \|u-v\|^2_{L^2(\Gamma)}$. Hence we obtain

$$\langle Au - Av, u-v \rangle \geq \|w_0\|^2_*.$$

Finally we get $\langle Au - Av, u-v \rangle \geq \frac{1}{c_{emb}^2}\|u-v\|^2_{H^{1/2}(\Gamma)}$.

(ii) boundedness
There holds

$$\langle \Phi(u), u \rangle \geq c_{min}\|u\|^2_{L^2(\Gamma)} + \langle \Phi(0), u \rangle$$
$$\geq c_{min}\|u\|^2_{L^2(\Gamma)} - \sqrt{|\Gamma|}\,|h(0)|\,\|u\|_{L^2(\Gamma)}$$

and hence

$$
\begin{aligned}
\langle Au, u \rangle &\geq \|w_0\|_*^2 - \sqrt{|\Gamma|}\, |h(0)|\, \|u\|_{L^2(\Gamma)} \\
&\geq \frac{1}{c_{emb}^2} \|u\|_{H^{1/2}(\Gamma)}^2 - \sqrt{|\Gamma|}\, |h(0)|\, \|u\|_{H^{1/2}(\Gamma)}
\end{aligned}
$$

where $w_0 \in H^1(\Omega)$ denotes the harmonic extension of $u \in H^{1/2}(\Gamma)$ to Ω. On the other hand we have

$$\langle g, u \rangle \leq \|g\|_{H^{-1/2}(\Gamma)} \|u\|_{H^{1/2}(\Gamma)}$$

which implies the assertion. $\square$

Remark

An example for a suitable h that satisfies (A2), is given by the truncation $\tilde{\alpha}$ of the heat transfer coefficient in (3.3). Thus (A2) is satisfied with $c_{min} = \frac{\alpha_l}{\lambda}$.

4.1.4. Iterative determination of the boundary temperature as a fixed point

We solve (4.7) iteratively. Hereto we propose a fixed point iteration based on Banach's fixed point theorem.

Using the notation from section 4.1.2 we define $\mathcal{T} : H^{1/2}(\Gamma) \to H^{1/2}(\Gamma)$ with $\mathcal{T}(u) := \left(\frac{I}{2} - \mathcal{K} + \mathcal{S} \circ \Phi\right)(u) - \mathcal{K}_b(f)$. The equation (4.7) for the boundary temperature u is satisfied iff the fixed point relation

$$\mathcal{G}_\gamma(u) := u - \gamma \mathcal{T}(u) = u \qquad (4.9)$$

holds for at least one $\gamma \in \mathbb{R} \setminus \{0\}$. By previous considerations there exists a unique fixed point $u \in H^{1/2}(\Gamma)$ for (4.9). Following the ideas in [13] and [11] we determine a γ which ensures that $\mathcal{G}_\gamma$ is a contraction in $H^{1/2}(\Gamma)$. For this purpose we need Lipschitz-continuity and strong monotonicity of $\mathcal{T}$ with respect to an appropriate norm in $H^{1/2}(\Gamma)$.

Equivalent norm in $H^{1/2}(\Gamma)$

By (A1), $\mathcal{S} : H^{-1/2}(\Gamma) \to H^{1/2}(\Gamma)$ is a strongly elliptic, self-adjoint operator and so is $\mathcal{S}^{-1} : H^{1/2}(\Gamma) \to H^{-1/2}(\Gamma)$. Thus the bilinear form $\langle u, v \rangle_{\mathcal{S}^{-1}(\Gamma)} := \langle u, \mathcal{S}^{-1}(v) \rangle_{L^2(\Gamma)}$ is symmetric. We introduce a norm on $H^{1/2}(\Gamma)$ induced by the inverse of the single layer operator

$$\|u\|_{\mathcal{S}^{-1}(\Gamma)}^2 := \langle u, \mathcal{S}^{-1}(u) \rangle_{L^2(\Gamma)}, \quad u \in H^{1/2}(\Gamma).$$

This norm is equivalent to the Sobolev-Slobodetskii-norm on $H^{1/2}(\Gamma)$, [44];

$$\exists c_\Gamma > 0 : \quad \frac{1}{c_\Gamma} \|u\|_{\mathcal{S}^{-1}(\Gamma)} \leq \|u\|_{H^{1/2}(\Gamma)} \leq c_\Gamma \|u\|_{\mathcal{S}^{-1}(\Gamma)}$$

Similar to the representation in (4.8), the main advantage of this equivalent norm is the separation of the nonlinearity Φ and the Single-layer operator $\mathcal{S}$ when $\|\cdot\|_{\mathcal{S}^{-1}(\Gamma)}$ is applied to $\mathcal{T}$. I.e. We are able to use the monotonicity assumption on Φ for the monotonicity of $\mathcal{T}$.

Now we want to establish conditions for Lipschitz-continuity of $\mathcal{T}$. To this end we inforce the assumption (A2) on the boundary map h by

(A2') *Monotonicity and Lipschitz continuity of h*
We require that $h : \mathbb{R} \to \mathbb{R}$ is Lipschitz-continuous satisfying $\exists c > 0 : |h(s_1) - h(s_2)| \leq c |s_1 - s_2|$ such that the respective superposition operator $\Phi(u)(x) := h(u(x))$, $\Phi : H^{1/2}(\Gamma) \to H^{-1/2}(\Gamma)$ is Lipschitz continuous. Moreover h shall satisfy the monotonicity estimate in (A2) such that Φ is strongly monotonous.

Lemma 4.2 (Lipschitz continuity of $\mathcal{T}$)
Suppose (A1) and (A2'). Then there exists $L > 0$ such that

$$\|\mathcal{T}(u) - \mathcal{T}(v)\|_{\mathcal{S}^{-1}(\Gamma)} \leq L \|u - v\|_{\mathcal{S}^{-1}(\Gamma)} \quad \text{for } u, v \in H^{1/2}(\Gamma).$$

Proof

By the definition of $\|\cdot\|_{\mathcal{S}^{-1}(\Gamma)}$ we identify $\|\mathcal{T}(u) - \mathcal{T}(v)\|^2_{\mathcal{S}^{-1}(\Gamma)}$ with

$$\left\langle \left(\frac{I}{2} - \mathcal{K}\right)(u-v) + \mathcal{S}(\varphi_u - \varphi_v),\ \mathcal{S}^{-1}\left(\left(\frac{I}{2} - \mathcal{K}\right)(u-v) + \mathcal{S}(\varphi_u - \varphi_v)\right) \right\rangle_{L^2(\Gamma)}$$

where φ_u denotes the image of the nonlinear superposition operator Φ, i.e. $\varphi_u = \Phi(u)$. $\mathcal{S}$ is self adjoint and we obtain

$$\begin{aligned}
\|\mathcal{T}(u) - \mathcal{T}(v)\|^2_{\mathcal{S}^{-1}(\Gamma)} &= \left\|\left(\frac{I}{2} - \mathcal{K}\right)(u-v)\right\|^2_{\mathcal{S}^{-1}(\Gamma)} + \|\mathcal{S}(\varphi_u - \varphi_v)\|^2_{\mathcal{S}^{-1}(\Gamma)} \\
&\quad + 2\left\langle \left(\frac{I}{2} - \mathcal{K}\right)(u-v),\ \mathcal{S}(\varphi_u - \varphi_v) \right\rangle_{\mathcal{S}^{-1}(\Gamma)} \\
&\leq \left\|\left(\frac{I}{2} - \mathcal{K}\right)(u-v)\right\|^2_{\mathcal{S}^{-1}(\Gamma)} + \|\mathcal{S}(\varphi_u - \varphi_v)\|^2_{\mathcal{S}^{-1}(\Gamma)} \\
&\quad + 2\left\|\left(\frac{I}{2} - \mathcal{K}\right)(u-v)\right\|_{\mathcal{S}^{-1}(\Gamma)} \|\mathcal{S}(\varphi_u - \varphi_v)\|_{\mathcal{S}^{-1}(\Gamma)}.
\end{aligned}$$

$\mathcal{S}: H^{-1/2}(\Gamma) \to H^{1/2}(\Gamma)$ and $\left(\frac{I}{2} - \mathcal{K}\right): H^{1/2}(\Gamma) \to H^{1/2}(\Gamma)$ are bounded linear operators and hence Lipschitz continuous. Thus we have

$$\begin{aligned}
\|\mathcal{T}(u) - \mathcal{T}(v)\|^2_{\mathcal{S}^{-1}(\Gamma)} &\leq L^2_{\frac{I}{2}-\mathcal{K}} \|u-v\|^2_{\mathcal{S}^{-1}(\Gamma)} + L^2_{\mathcal{S}} \|\varphi_u - \varphi_v\|^2_{H^{-1/2}(\Gamma)} \\
&\quad + 2 L_{\frac{I}{2}-\mathcal{K}} L_{\mathcal{S}} \|u-v\|_{\mathcal{S}^{-1}(\Gamma)} \|\varphi_u - \varphi_v\|_{H^{-1/2}(\Gamma)}.
\end{aligned}$$

Now the Lipschitz continuity of Φ implies

$$\|\mathcal{T}(u) - \mathcal{T}(v)\|^2_{\mathcal{S}^{-1}(\Gamma)} \leq \left(L^2_{\frac{I}{2}-\mathcal{K}} + 2 L_{\frac{I}{2}-\mathcal{K}} L_{\mathcal{S}} L_\Phi + L^2_{\mathcal{S}} L^2_\Phi \right) \|u-v\|^2_{\mathcal{S}^{-1}(\Gamma)}.$$

where L_Φ denotes $\|\Phi(u) - \Phi(v)\|_{H^{-1/2}(\Gamma)} \leq L_\Phi \|u-v\|_{\mathcal{S}^{-1}(\Gamma)}$. Thus we get the assertion with $L = L_{\frac{I}{2}-\mathcal{K}} + L_{\mathcal{S}} L_\Phi$. $\square$

Lemma 4.3 (Strong monotonicity of $\mathcal{T}$)
Suppose (A1) and (A2'). Then there exists $m > 0$ such that

$$\left\langle \mathcal{T}(u) - \mathcal{T}(v),\ \mathcal{S}^{-1}(u-v) \right\rangle_{L^2(\Gamma)} \geq m \|u-v\|^2_{\mathcal{S}^{-1}(\Gamma)} \quad \text{for } u, v \in H^{1/2}(\Gamma).$$

Proof

$\mathcal{S}^{-1}$ is self adjoint, hence we have

$$\langle \mathcal{T}(u) - \mathcal{T}(v), \mathcal{S}^{-1}(u-v) \rangle_{L^2(\Gamma)} = \langle Au - Av, u-v \rangle_{L^2(\Gamma)}$$

where $A = \mathcal{P} + \Phi$ and $\mathcal{P}$ denotes the Steklov-Poincaré Operator defined in section 4.1.3. As in the proof of Theorem 4.1 we have

$$\langle Au - Av, u-v \rangle_{L^2(\Gamma)} \geq \frac{1}{c_{emb}^2} \|u-v\|_{H^{1/2}(\Gamma)}^2$$

where c_{emb} denotes the constant of the trace embedding between $H^{1/2}(\Gamma)$ and $H^1(\Omega)$; w.r.t. $\|\cdot\|_\star$. Hence we have

$$\langle \mathcal{T}(u) - \mathcal{T}(v), \mathcal{S}^{-1}(u-v) \rangle_{L^2(\Gamma)} \geq \frac{1}{c_{emb}^2 \, c_\Gamma^2} \|u-v\|_{\mathcal{S}^{-1}(\Gamma)}^2$$

which yields the assertion. $\square$

Construction of the iterative sequence

Now we establish an iterative sequence $(u^{(n)})_{n \in \mathbb{N}} \subset H^{1/2}(\Gamma)$ which converges to the solution of (4.7) for an arbitrary initial function $u^{(1)} \in H^{1/2}(\Gamma)$.

Theorem 4.2

Let the assumptions (A1) and (A2') hold. Define the iterative sequence $(u^{(n)})_{n \in \mathbb{N}} \subset H^{1/2}(\Gamma)$ by $u^{(n+1)} := \mathcal{G}_\gamma(u^{(n)})$, $\gamma = m/L^2$ where L denotes the $\mathcal{S}^{-1}$- Lipschitz constant and m the $\mathcal{S}^{-1}$-monotonicity constant of $\mathcal{T}$. Then, for every initial function $u^{(1)} \in H^{1/2}(\Gamma)$, $(u^{(n)})_{n \in \mathbb{N}}$ converges to the solution u of (4.7) with respect to $\|\cdot\|_{\mathcal{S}^{-1}}$ with the a priori error estimate

$$\|u^{(n)} - u\|_{\mathcal{S}^{-1}(\Gamma)} \leq \frac{k^n}{1-k} \|u^{(2)} - u^{(1)}\|_{\mathcal{S}^{-1}(\Gamma)}, \quad k = \sqrt{1 - \frac{m^2}{L^2}}.$$

Proof

It suffices to verify that $\mathcal{G}_\gamma : H^{1/2}(\Gamma) \to H^{1/2}(\Gamma)$ is contractive w.r.t.

$\|\cdot\|_{\mathcal{S}^{-1}(\Gamma)}$. Then the assertions of Theorem 4.2 follow by Banach's fixed point theorem.

$$\begin{aligned}\|\mathcal{G}_\gamma(u) - \mathcal{G}_\gamma(v)\|^2_{\mathcal{S}^{-1}(\Gamma)} &= \|u-v\|^2_{\mathcal{S}^{-1}(\Gamma)} + \gamma^2 \|\mathcal{T}(u) - \mathcal{T}(v)\|^2_{\mathcal{S}^{-1}(\Gamma)} \\ &\quad - 2\gamma \langle \mathcal{T}(u) - \mathcal{T}(v), \mathcal{S}^{-1}(u-v)\rangle_{L^2(\Gamma)} \\ &\leq (1 - 2m\gamma + L^2\gamma^2)\|u-v\|^2_{\mathcal{S}^{-1}(\Gamma)}\end{aligned}$$

The estimate is provided by the Lipschitz continuity and strong monotonicity of $\mathcal{T}$. The minimum of $1 - 2m\gamma + L^2\gamma^2$ is attained at $\gamma = \frac{m}{L^2}$ and amounts to $1 - \frac{m^2}{L^2}$. Hence we get $k = \sqrt{1 - \frac{m^2}{L^2}}$ as the constant of contraction. □

4.1.5. Case of rotational symmetry

In this section we consider problem (4.2) in the rotationally symmetric domain $\Omega = B_r(0) \subset \mathbb{R}^2$ to obtain a benchmark for the boundary integral formulation. First we identify the associated

Boundary integral operators on arcs of circles

For constant Poisson-data f we compute the Bi-Laplace double layer operator

$$(\mathcal{K}_b f)(x) = -f \int_\Gamma \frac{\partial}{\partial n_y} F_b(x-y)\, \mathrm{d}s_y \quad \text{where}$$

$$\begin{aligned}\frac{\partial}{\partial n_y}F_b(x-y) &= \langle \nabla F_b(x-y), n_y\rangle_{\mathbb{R}^2} = \left\langle \frac{y-x}{4\pi}\left(\ln|x-y| - \frac{1}{2}\right), \frac{y}{|y|}\right\rangle_{\mathbb{R}^2} \\ &= \left(\frac{\ln|x-y|}{4\pi} - \frac{1}{8\pi}\right)\left(\frac{|y|^2 - \langle x, y\rangle_{\mathbb{R}^2}}{|y|}\right).\end{aligned}$$

We parametrize $\Gamma = \partial B_r(0)$ via $\gamma : [0, 2\pi] \to \Gamma$ with $y = \gamma(t) = r\begin{pmatrix}\cos t \\ \sin t\end{pmatrix}$ and set $x = r\begin{pmatrix}\cos t_0 \\ \sin t_0\end{pmatrix}$ for some fixed $t_0 \in [0, 2\pi]$. I.e.

$$\frac{\partial}{\partial n_y}F_b(x-y) = \frac{r}{4\pi}\left(\ln(2r) + \ln(1 - \cos(t_0 - t)) - \frac{1}{2}\right)(1 - \cos(t_0 - t)).$$

Thus the parametrization yields

$$(\mathcal{K}_b f)(x) = f \frac{r^2}{4\pi}\left(\frac{1}{2} - \ln(2r)\right) \int_0^{2\pi} (1 - \cos(t_0 - t))\, dt$$

$$- f \frac{r^2}{4\pi} \int_0^{2\pi} \ln(1 - \cos(t_0 - t))(1 - \cos(t_0 - t))\, dt$$

$$= -f \frac{r^2}{2}\left(\frac{1}{2} + \ln(r)\right)$$

It remains to compute the single layer operator
$(\mathcal{S}\varphi)(x) = -\int_\Gamma \varphi(y) F(x-y)\, ds_y$ and the double layer operator
$(\mathcal{K}u)(x) = \int_\Gamma u(y) \frac{\partial}{\partial n_y} F(x-y)\, ds_y$ for constant Neumann and Dirichlet data φ
and u. We have $F(x-y) = \frac{1}{2\pi}\ln(|x-y|)$ and

$$\frac{\partial}{\partial n_y} F(x-y) = \frac{1}{2\pi |x-y|^2}\left(\frac{|y|^2 - \langle x, y\rangle_{\mathbb{R}^2}}{|y|}\right)$$

Thus - using the parametrization $\gamma = \gamma(t)$ from above - there holds

$$\int_{\partial B_r} \frac{\partial}{\partial n_y} F(x-y)\, ds_y = \frac{1}{2} \quad \text{and} \quad \int_{\partial B_r} F(x-y)\, ds_x = r\ln r.$$

This yields

$$(\mathcal{S}\varphi)(x) = -\varphi r \ln r \quad \text{and} \quad (\mathcal{K}u)(x) = \frac{u}{2}.$$

Equation (4.7) and iterative determination of the boundary temperature in the case of rotational symmetry

Using the rotationally symmetric representations (4.7) reads as

$$-r \ln r\, \Phi(u) + \frac{r^2}{2}\left(\frac{1}{2} + \ln(r)\right) f = 0. \tag{4.10}$$

The solution to this equation can be found directly via Newton's method or via a fixed point iteration.

We illustrate the latter method in an application to physical quantities for electric cables. Using the notation of section 4.1.1 we identify

$$\Phi(u) = \frac{\alpha(u)}{\lambda}(u - u_{env}) \quad \text{and} \quad f = \frac{\rho_0 \left(1 + \alpha_\rho u_m\right) I^2}{\lambda |\Omega|^2}.$$

Thus equation (4.10) reads as

$$u = u_{env} + \underbrace{\left(1 + \frac{1}{2\ln(r)}\right)}_{a_r} \frac{\rho_0 \left(1 + \alpha_\rho u_m\right) I^2}{2\pi^2 r^3 \alpha(u)} =: \zeta_b(u) \qquad (4.11)$$

A priori determination of the mean value u_m

Here we use the energy conservating mean value u_m to evaluate the resistivity $\rho = \rho(u_m) = \rho_0 \left(1 + \alpha_\rho u_m\right)$. It is determined a priori as proposed in section 4.1.1 via a fixed point iteration of the equation

$$u_m = u_{env} + \frac{\rho_0 \left(1 + \alpha_\rho u_m\right) I^2}{2\pi^2 r^3 \alpha(u_m)}. \qquad (4.12)$$

The iteration is contractive - i.e. converges globally - provided the relation $\alpha_\rho < \frac{2\pi^2 r^3 \alpha_l}{\rho_0 I^2}$ holds. For this u_m we can now solve (4.11) via a fixed point iteration.

Proposition 4.2

Let $\alpha : [u_{env}, \infty) \to [\alpha_l, \alpha_h]$; *denote the truncated heat transfer coefficient from (3.3) and let the truncation yield a differentiable map α such that $\alpha(u) \geq \alpha'(u) u_{env}$ holds for $u \in [u_{env}, \infty)$. Moreover let the relation $a_r \rho_0 \left(1 + \alpha_\rho u_m\right) I^2 < 2\pi^2 r^3 \alpha_l u_{env}$ hold. Define the iterative sequence $(u^{(n)})_{n \in \mathbb{N}} \subset \mathbb{R}$ by $u^{(n+1)} := \zeta_b(u^{(n)})$. Then, for any initial value $u^{(1)} \in [u_{env}, \infty)$ the iterative sequence $(u^{(n)})_{n \in \mathbb{N}}$ converges to the unique solution u of (4.12) with the following rate of convergence*

$$\left|u^{(n)} - u\right| \leq \frac{q^n}{1-q}\left|u^{(2)} - u^{(1)}\right| \quad \text{where} \quad q := \frac{\rho_0 \left(1 + \alpha_\rho u_m\right) I^2 a_r}{\alpha_l u_{env} 2\pi^2 r^3}.$$

Proof

We show that $a_r \, \rho_0 \, (1 + \alpha_\rho) \, I^2 < 2 \, \pi^2 \, r^3 \, \alpha_l \, u_{env}$ yields global contractivity of the map $\zeta_b : [u_{env}, \infty) \to [u_{env}, \infty)$. There holds

$$\sup_{s \in [u_{env}, \infty)} |\zeta_b'(s)| = \sup_{s \in [u_{env}, \infty)} \frac{\rho_0 \, (1 + \alpha_\rho \, u_m) \, I^2 \, a_r}{2 \, \pi^2 \, r^3} \left| \frac{\alpha'(s)}{\alpha^2(s)} \right|$$

$$\leq \frac{\rho_0 \, (1 + \alpha_\rho \, u_m) \, I^2 \, c_r}{\alpha_l \, u_{env} \, 2 \, \pi^2 \, r^3} = q < 1 \, .$$

This implies $|\zeta_b(s_2) - \zeta_v(s_1)| \leq q \, |s_2 - s_1|$ for $s_1, s_2 \in [v_{env}, \infty)$. Thus existence and uniqueness of a solution of (4.12) and convergence of the iterative $(u_n)_{n \in \mathbb{N}}$ sequence follow by Banach's fixed point theorem □

An example for a heat transfer coefficient which fulfills the requirements of Proposition 4.2 is given in section 3.1.5.

Application to physical data

As in section 3.1.5 we fix the physical data with $r = 2 * 10^{-3}$, $\rho_0 = 1.72 * 10^{-8}$, $\alpha_\rho = 3.83 * 10^{-3}$, $u_l = u_{env} = 25$, $\alpha_l = 10$.

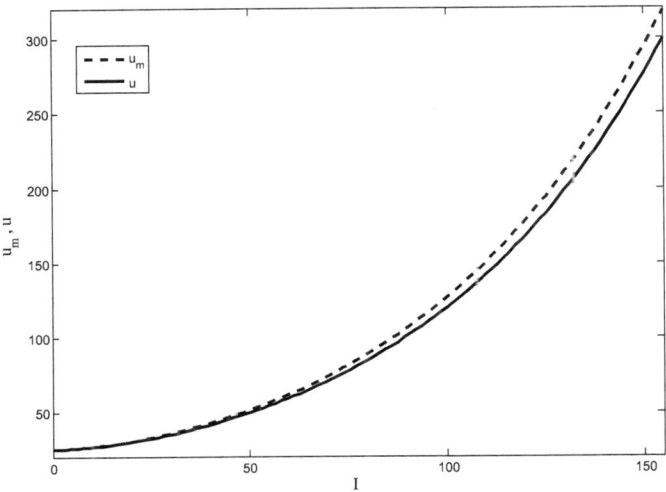

First we determine u_m depending on the current $I \in [0, 154.8]$. To this data we then solve (4.11). The dependence of the mean value temperature u_m and the boundary temperature u on I is depicted in the preceding graph.

Relation between u and u_m

As one shall expect, the application to physical data shows that the boundary temperature u is smaller than the mean value temperature u_m. We fix this observation in the following.

Proposition 4.3
Let u_m, u denote the solutions of (4.11), (4.12) respectively, and suppose $0 < r < \frac{1}{\sqrt{e}}$. Then there holds

$$u - u_{env} \leq u_m - u_{env} \leq \frac{1}{a_r}(u - u_{env})$$

where $a_r = 1 + \frac{1}{2 \ln r} < 1$.

Proof

We argue by contradiction. Assume $u > u_m$. This implies $\frac{u - u_{env}}{u_m - u_{env}} > 1$. On the other hand the Equations (4.11) and (4.12) yield

$$\frac{u - u_{env}}{u_m - u_{env}} = a_r \frac{\alpha(u_m)}{\alpha(u)}.$$

Since we have $\left(1 + \frac{1}{2 \ln r}\right) < 1$, the monotonicity of α gives $\frac{u - u_{env}}{u_m - u_{env}} < 1$, a contradiction.

Thus there holds $u \leq u_m$ which provides the first inequality in Proposition 4.3. The second one follows by $\frac{u_m - u_{env}}{u - u_{env}} = \frac{1}{a_r} \frac{\alpha(u)}{\alpha(u_m)} \leq \frac{1}{a_r}$. $\square$

The example used in the application to physical data, i.e. $r = 2 * 10^{-3}$ gives

$$u - u_{env} \leq u_m - u_{env} \leq 1.088\, (u - u_{env}).$$

4.1.6. Illustration of Theorem 4.2

The direct fixed point iteration presented above is the most elementary and thus preferable method to solve (4.10).

Nevertheless we present how Theorem 4.2 and the corresponding setting apply to the case of rotational symmetry. It shall serve as a benchmark for non-symmetric domains. First we identify

$$\mathcal{T}(u) = \left(\frac{I}{2} - \mathcal{K} + \mathcal{S} \circ \Phi\right)(u) - \mathcal{K}_b(f) = -r \ln r\, \Phi(u) + \frac{r^2}{2}\left(\frac{1}{2} + \ln(r)\right) f.$$

Next we determine the Lipschitz and monotonicity constant L, m of $\mathcal{T}$ w.r.t. $\|\cdot\|_{\mathcal{S}^{-1}(\Gamma)}$, $\Gamma = \partial B_r$.

Lipschitz estimate

Since u is constant on ∂B_r we have

$$\|\mathcal{T}(u) - \mathcal{T}(v)\|_{\mathcal{S}^{-1}(\Gamma)} = \|1\|_{\mathcal{S}^{-1}(\Gamma)} |\mathcal{T}(u) - \mathcal{T}(v)|$$

where $\|1\|_{\mathcal{S}^{-1}(\Gamma)} = \sqrt{\int_\Gamma \mathcal{S}^{-1}(1)\,ds} = \sqrt{\frac{2\pi}{-\ln r}}$, $r < 1$. This implies

$$\|\mathcal{T}(u) - \mathcal{T}(v)\|_{\mathcal{S}^{-1}(\Gamma)} = -r \ln r \sqrt{\frac{2\pi}{-\ln r}}\, |\Phi(u) - \Phi(v)|$$

where $\Phi(u) = \frac{\alpha(u)}{\lambda}(u - u_{env})$. Considering a truncation of α which fulfills $\alpha'(u)(u - u_{env}) \leq \alpha_h$, $u \in [u_{env}, \infty]$ we get

$$|\Phi(u) - \Phi(v)| \leq \sup_{s \in [u_{env}, \infty)} |\Phi'(s)|\, |u - v|$$

$$= \frac{1}{\lambda} \sup_{s \in [u_{env}, \infty)} (\alpha'(s)(s - u_{env}) + \alpha(s))\, |u - v| \leq \frac{2\alpha_h}{\lambda} |u - v|.$$

This yields

$$\|\mathcal{T}(u) - \mathcal{T}(v)\|_{\mathcal{S}^{-1}(\Gamma)} \leq \frac{-2r \ln r\, \alpha_h}{\lambda} \|u - v\|_{\mathcal{S}^{-1}(\Gamma)} \quad \text{i.e.} \quad L = \frac{-2r \ln r\, \alpha_h}{\lambda}.$$

Monotonicity estimate
There holds

$$\langle \mathcal{T}(u) - \mathcal{T}(v), \mathcal{S}^{-1}(u-v) \rangle = \int_\Gamma -r \ln r \, (\Phi(u) - \Phi(v)) \frac{1}{-r \ln r} (u-v) \, ds$$
$$= 2\pi r \, (\Phi(u) - \Phi(v)) (u-v)$$
$$\geq \frac{2\pi r \alpha_l}{\lambda} (u-v)^2 = -\frac{r \ln r \alpha_l}{\lambda} \|u-v\|^2_{\mathcal{S}^{-1}(\Gamma)}$$

i.e. $m = -\frac{r \ln r \alpha_l}{\lambda}$.

Construction of the iterative sequence
Setting $\mathcal{G}_\gamma(u) = u - \gamma \mathcal{T}(u)$ with $\gamma = \frac{m}{L^2} = -\frac{\lambda \alpha_l}{4r \ln r \alpha_h^2}$ we obtain the iteratively defined sequence $u^{(n+1)} = \mathcal{G}_\gamma(u^{(n)})$ which converges for any initial value $u^{(1)} \in [u_{env}, \infty)$ by Theorem 4.2. We have

$$\mathcal{G}_\gamma(u) = u + \frac{\lambda \alpha_l}{4r \ln r \alpha_h^2} \left(-r \ln r \, \Phi(u) + \frac{r^2}{2} \left(\frac{1}{2} + \ln(r) \right) f \right)$$

with $\Phi(u) = \frac{\alpha(u)}{\lambda}(u - u_{env})$ and $f = \frac{\rho_0 (1 + \alpha_\rho u_m) I^2}{\lambda (\pi r^2)^2}$ and hence

$$\mathcal{G}_\gamma(u) = u + \frac{\alpha_l}{4\alpha_h^2} \left(\left(1 + \frac{1}{2 \ln r}\right) \frac{\rho_0 (1 + \alpha_\rho u_m) I^2}{2\pi^2 r^3} - \alpha(u)(u - u_{env}) \right).$$

Observe that a fixed point iteration of $\mathcal{G}_\gamma$ yields the same limit as the iteration of ζ_b in (4.11). Moreover the iteratively defined sequence $(u^{(n)})_{n \in \mathbb{N}}$ converges with $k = \sqrt{1 - \frac{m^2}{L^2}} = \sqrt{1 - \frac{\alpha_l^2}{4\alpha_h^2}}$ w.r.t. $\|\cdot\|_{\mathcal{S}^{-1}(\Gamma)}$.

Remarks

(i) Our fixed point approach can be applied to non-symmetric domains and result in numerical methods, provided the fixed point iteration is combined with numerical quadrature for the occuring singular integrals. We refer to [1], [21], [22], [73].

(ii) Another classical iterative approach for (4.7) is Newton's method for nonlinear boundary integral equations. It is investigated in combination with the Galerkin boundary element method in [34] and [35]. The observed efficiency of this method has a drawback in the implicit character of the convergence conditions. In particular, the initial iterative step to be chosen in an a priori unknown neighbourhood of the solution. Additionally one needs the Fréchet-Calculus for integral operators - see e.g. [68] - which characterizes the derivative in Newton's method.

The rate of convergence of the fixed point approach may be worse than in Newton's method. On the other hand, we are able to give an explicit a priori analysis for global convergence, using elementary iteration steps.

4.2. Boundary integral approach for insulated cables

4.2.1. Setup of the problem

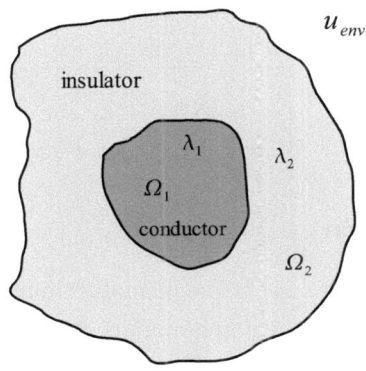

Analogously to section 3.2 we describe the cross section of the insulated cable by the bounded, simply connected and open union $\Omega_{cr} = \overline{\Omega}_1 \cup \Omega_2 \subset \mathbb{R}^2$ with Lipschitz boundaries $\partial \Omega_{cr}$, $\partial \Omega_1$.

The stationary temperature distribution $u_{st} : \Omega_{cr} \to \mathbb{R}$ has to satisfy the following boundary value problem

$$-\lambda_1 \Delta u_{st} = f(u_{st}) \quad \text{in } \Omega_1 \qquad (4.13)$$
$$-\lambda_2 \Delta u_{st} = 0 \quad \text{in } \Omega_2 \qquad (4.14)$$
$$-\lambda_2 \frac{\partial u_{st}}{\partial n} = \alpha(u_{st})(u_{st} - u_{env}) \quad \text{on } \partial\Omega_{cr}.$$

Restriction of the temperature dependence of f to $\bar{u}$

We approximate (4.13; 4.14) by

$$-\lambda_1 \Delta u = f(\bar{u}) \quad \text{in } \Omega_1 ; \quad -\lambda_2 \Delta u = 0 \quad \text{in } \Omega_2 \qquad (4.15)$$
$$-\lambda_2 \frac{\partial u}{\partial n} = \alpha(u)(u - u_{env}) \quad \text{on } \partial\Omega_{cr}.$$

where $f(\bar{u}) = \frac{\rho_0 (1+\alpha_\rho \bar{u}) I^2}{|\Omega_1|^2}$ for some $\bar{u} \in \mathbb{R}$.

For the following asymptotic result we use the transformation of the monotone boundary condition for insulated cables from section 3.2.3 and its application in section 3.2.4. Thus existence and uniqueness for (4.13; 4.14) combined with an error estimate for the approximation by (4.15) read as

Assume $\alpha_\rho < \frac{\lambda_1 |\Omega_1|^2}{\rho_0 I^2 c_\star^2}$. Then there exists a unique solution $u_{st} \in H^1(\Omega_1)$ of (4.13; 4.14) which is approximated by the solution of (4.15) via

$$\|u_{st} - u\|_\star \leq C_{\alpha_\rho} \|u - \bar{u}\|_{L^2(\Omega_1)} \quad \text{where} \quad C_{\alpha_\rho} = \frac{|\alpha_\rho| \rho_0 I^2 c_\star}{\lambda |\Omega_1|^2 - |\alpha_\rho| \rho_0 I^2 c_\star^2}.$$

4.2.2. The outer domain formulation

Now we formulate the boundary value problem (4.15) in the insulator domain Ω_2 only. To this end we choose the constant mean value boundary temperature (m.v.b.t.) $\bar{u} := \frac{1}{|\partial\Omega_1|} \int_{\partial\Omega_1} u \, d\sigma$ in the Poisson datum of (4.15). This is not the error minimizing choice; nevertheless, as we shall see, it is the appropriate one for the forthcoming boundary integral formulation.

Consider now the heat flow density $q = q(u)$ over the boundary $\partial\Omega_1$ which enters in the inner boundary condition $-\lambda_1 \frac{\partial u_{cond}}{\partial n} = q$. Using the equality of heat flows $\lambda_1 \frac{\partial u_{cond}}{\partial n} = \lambda_2 \frac{\partial u_{ins}}{\partial n}$ this condition becomes $-\lambda_2 \frac{\partial u}{\partial n} = q$ on $\partial \Omega_1$ for $u = u_{ins}$. Assume for the moment the heat flow density q as given. Note that by the Divergence Theorem, q has to fulfill

$$\int_{\partial \Omega_1} q \, d\sigma = |\Omega_1| f(\bar u) = \rho(\bar u) \frac{I^2}{|\Omega_1|}. \qquad (4.16)$$

The simplified form of the right hand side is justified by the approximation estimate above. Thus we consider the following boundary value problem

$$-\Delta u = 0 \quad \text{in } \Omega_2 =: \Omega \qquad (4.17)$$

$$\lambda_2 \frac{\partial u}{\partial n} = q(u) \quad \text{on } \partial \Omega_1 =: \Gamma_1 \qquad (4.18)$$

$$-\lambda_2 \frac{\partial u}{\partial n} = \alpha(u)(u - u_{env}) \quad \text{on } \partial \Omega \setminus \partial \Omega_1 =: \Gamma_2. \qquad (4.19)$$

4.2.3. Determination of the heat flow

For the computation of $q = q(u)$ one has to regard the specific geometry of the boundary Γ_1 and the source term $f = f(u)$. The general situation can be treated as an inverse problem. We refer to [8], [28], [32], [78].

Dual mixed formulations

Another possibility is given by the dual mixed formulation of (4.13,4.14). Here we search for a pair $(\mathbf{q}, u)$ of solutions where $\mathbf{q} = \nabla u$ denotes the heat flow. First we define the extension of f via $\bar f = \frac{1}{\lambda_1} f(u) \mathbb{I}_{\Omega_1}(x)$, $x \in \Omega_{cr}$ and recall the variational formulation of (4.13,4.14) which reads as

$$\int_{\Gamma_2} \beta(u) v \, d\sigma + \int_{\Omega_{cr}} \nabla u \nabla v \, dx = \int_{\Omega_{cr}} \bar f(u) \, dx, \; \forall v \in H^1(\Omega_{cr})$$

where $\beta(u) = \frac{1}{\lambda_2} \alpha(u)(u - u_{env})$. Due to the discussed properties of α (section 3.1.5) the associated superposition operator $\mathcal{B}(u)(x) = \beta(u(x))$ is strongly monotone and maps $L^2(\Gamma_2)$ into $L^2(\Gamma_2)$. Hence the dual mixed formulation reads as: Find $(\mathbf{q}, u) \in L^2(\Omega_{cr})^2 \times H^1(\Omega_{cr})$ such that

$$\begin{aligned}\langle \mathcal{B}(u), v\rangle_{L^2(\Gamma_2)} + \langle \mathbf{q}, \nabla v\rangle &= \langle \bar{f}(u), v\rangle_{L^2(\Omega_{cr})}, \quad \forall v \in H^1(\Omega_{cr}) \\ \langle \mathbf{q}, \tau\rangle &= \langle \nabla u, \tau\rangle, \quad \forall \tau \in L^2(\Omega_{cr})^2 \end{aligned} \quad (4.20)$$

where $\langle \cdot, \cdot \rangle = \langle \cdot, \cdot \rangle_{L^2(\Omega_{cr})^2}$. Here we recover the saddle point form of (4.20)

$$\begin{pmatrix} \mathcal{B} & \nabla^* \\ -\nabla & \mathbb{I} \end{pmatrix} \begin{pmatrix} u \\ \mathbf{q} \end{pmatrix} = \begin{pmatrix} \bar{f}(u) \\ 0 \end{pmatrix} \quad (4.21)$$

where $\mathbb{I} : L^2(\Omega_{cr})^2 \to L^2(\Omega_{cr})^2$ denotes the identity map and $\nabla^* : L^2(\Omega_{cr})^2 \to (H^1(\Omega_{cr}))^*$ is a formal definition of the adjoint of $\nabla : H^1(\Omega_{cr}) \to L^2(\Omega_{cr})^2$,

$$\langle \nabla^* \mathbf{q}, v\rangle_{L^2(\Omega_{cr})} = \langle \mathbf{q}, \nabla v\rangle_{(L^2(\Omega_{cr}))^2} \quad \forall (\mathbf{q}, v) \in L^2(\Omega_{cr})^2 \times H^1(\Omega_{cr})$$

On the other hand the divergence theorem yields an alternative mixed formulation which imposes higher regularity on the flow $\mathbf{q}$ and a less regular u. Hereto we define

$$H(\text{div}, \Omega) = \{\tau \in L^2(\Omega)^2 \,;\, \text{div}\,\tau \in L^2(\Omega)\}$$

and the associated mixed form reads as: Find $(\mathbf{q}, u) \in H(\text{div}, \Omega_{cr}) \times L^2(\Omega_{cr})$ such that

$$\begin{aligned} -\langle \text{div}\,\mathbf{q}, v\rangle &= \langle \bar{f}(u), v\rangle, \quad \forall v \in L^2(\Omega_{cr}) \\ \langle \mathbf{q}, \tau\rangle_{L^2(\Omega_{cr})^2} &= \langle u, \text{div}\,\tau\rangle - \langle \mathcal{B}^{-1}(\mathbf{q}\cdot n), \tau\cdot n\rangle_{L^2(\Gamma_2)}, \quad \forall \tau \in H(\text{div}, \Omega_{cr}) \end{aligned} \quad (4.22)$$

where $\langle \cdot, \cdot\rangle = \langle \cdot, \cdot\rangle_{L^2(\Omega_{cr})}$.

For a numerical treatment of (4.20,4.22) in the linear case we refer to [10]. An existence result and a numerical treatment of (4.21) with stronger assumptions is provided by [36]. Nevertheless, the numerical analysis of (4.20,4.22) remains an open problem.

Coupled formulation

An effective numerical method solving (4.13,4.14) is FEM-BEM coupling between Ω_1 and Ω_2. In order to make (4.13,4.14) accessible for such a method one can use the following setting.

$$-\lambda_1 \Delta u_1 = f(u_1) \quad \text{in } \Omega_1 \qquad (4.23)$$
$$-\lambda_2 \Delta u_2 = 0 \quad \text{in } \Omega_2 \qquad (4.24)$$
$$-\lambda_2 \frac{\partial u_2}{\partial n} = \alpha(u_2)(u_2 - u_{env}) \quad \text{on } \Gamma_2.$$

The problems (4.23,4.24) are coupled by the following transmission condition on Γ_1:
$$u_1 = u_2 \quad \text{and} \quad \lambda_1 \frac{\partial u_1}{\partial n} = \lambda_2 \frac{\partial u_2}{\partial n}$$

i.e. continuity of the temperature and equality of the heat flows. Now it is possible to treat (4.23) via a mixed dual formulation with $\mathbf{q}_i = \nabla u_i$; $i = 1, 2$ and $\langle \cdot, \cdot \rangle = \langle \cdot, \cdot \rangle_{L^2(\Omega_1)^2}$ as sketched above;

$$\langle \lambda_2\, \mathbf{q}_2 \cdot n, v \rangle_{L^2(\Gamma_1)} + \langle \lambda_1\, \mathbf{q}_1, \nabla v \rangle = \langle f(u_1), v \rangle_{L^2(\Omega_1)}, \quad \forall v \in H^1(\Omega_1)$$
$$\langle \mathbf{q}_1, \tau \rangle = \langle \nabla u_1, \tau \rangle, \quad \forall \tau \in L^2(\Omega_1)^2.$$

(4.24) can be handled with the help of the fundamental solution of the Laplacian via a boundary integral approach. We will follow this approach for the uncoupled problem (4.17) in the next sections. A FEM-BEM method for the linear case is presented in [37], while a treatment of (4.23,4.24) via a FEM-BEM coupling is outstanding.

Approximation by rotational symmetry

In our case the source term is given by the m.v.b.t. approximation discussed above and reads as $f = f(\bar{u}) = \frac{\rho(\bar{u})\, I^2}{|\Omega_1|^2}$. Now we suggest an explicit form of the heat flow density for the following considerations. Since conductor cross sections of electric cables are nearly rotationally symmetric, let us assume $q = q(\bar{u})$. I.e. q does not depend on $x \in \Gamma_1$. Then (4.16) yields $q = q(\bar{u}) = \frac{\rho(\bar{u}) I^2}{|\partial \Omega_1| |\Omega_1|}$. Now if we drop the assumption that u is constant then, by (4.16), q and f have locally the same monotonicity behaviour w.r.t. to the boundary temperature. Thus we approximate a temperature dependent heat flux by

$$\tilde{q}(u) := \frac{\rho(u) I^2}{|\Gamma_1| |\Omega_1|}.$$

We observe that by the weak maximum principle (see e.g. [40]) the extremal values of u are attained at the boundary of Ω. In applications, these values are the most interesting ones which motivates the

4.2.4. Boundary integral approach on doubly connected domains

In the following we are concerned with the temperatures on the boundary of the outer domain Ω only. Using Green's representation formula we derive an equivalent nonlinear boundary integral equation for the doubly connected domain Ω with $\partial \Omega := \Gamma = \Gamma_1 \cup \Gamma_2$ that includes the boundary conditions (4.18), (4.19). Starting from $-\Delta u = 0$ in Ω the representation formula for harmonic functions and the jump relations of potential theory yield for the boundary values of u:

$$u(x) = 2 \int_\Gamma \left(u(y) \frac{\partial}{\partial n_y} F(x-y) - \frac{\partial u(y)}{\partial n_y} F(x-y) \right) ds_y, \quad x \in \Gamma \quad (4.25)$$

As before $F(z) := \frac{1}{2\pi} \ln(|z|)$ denotes the fundamental solution of the Laplace-equation in $\mathbb{R}^2 \setminus \{0\}$.

Here it is not necessary to consider a Newton potential and its transformation to the boundary via the Bi-Laplace double layer operator $\mathcal{K}_b$ such as in section 4.1.2. The outer domain Ω is source free and the heat source is given by the flux q in (4.18).

We introduce the following notation. $u_i := u|_{\Gamma_i}$; $i = 1, 2$ for the boundary temperatures and $\varphi_i = -\frac{\partial u_i}{\partial n}$ on Γ_i for the associated heat flux. Let $h : \mathbb{R}^2 \to \mathbb{R}^2$ be defined componentwise by

$$h(s_1, s_2) = \frac{1}{\lambda_2} \begin{pmatrix} -q(s_1) \\ \alpha(s_2)(s_2 - u_{env}) \end{pmatrix}.$$

Thus we get the superposition operator
$\Phi(u)(x_1, x_2) := h(u_1(x_1), u_2(x_2))$, $x_i \in \Gamma_i$ with the mapping property
$\Phi : H^{1/2}(\Gamma) \to H^{-1/2}(\Gamma)$. There holds $\begin{pmatrix} \varphi_1 \\ \varphi_2 \end{pmatrix} = \Phi(u)$.

We emphasize that we consider a heat flux $q = q(u_1)$ that may fully depend on the boundary temperature which can be obtained by an inverse treatment or experimental data. Note that the nonlinearity of Φ appears in the second component, due to the heat transfer coefficient $\alpha = \alpha(u_2)$, that enters in the outer boundary condition.

The function spaces for the boundary Γ of the doubly connected domain Ω are given by $H^s(\Gamma) := H^s(\Gamma_1) \times H^s(\Gamma_2)$, $\|\cdot\|^2_{H^s(\Gamma)} := \|\cdot\|^2_{H^s(\Gamma_1)} + \|\cdot\|^2_{H^s(\Gamma_2)}$, $\langle u, v \rangle_{H^s(\Gamma)} = \langle u_1, v_1 \rangle_{H^s(\Gamma_1)} + \langle u_2, v_2 \rangle_{H^s(\Gamma_2)}$
for $s \in \{-1/2, 1/2\}$; see e.g. [6], [41], [45] for various approaches in multiply connected domains.

Representation by single and double layer potential operators in a doubly connected domain

We define the following continuous mappings: The single layer potential operator $\mathcal{S} : H^{-1/2}(\Gamma) \to H^{1/2}(\Gamma)$ by

$$\mathcal{S}(\varphi)(x) = -\int_\Gamma \varphi(y) F(x-y) \, ds_y = \begin{pmatrix} S_{11} & S_{12} \\ S_{21} & S_{22} \end{pmatrix} \begin{pmatrix} \varphi_1 \\ \varphi_2 \end{pmatrix} \quad (4.26)$$

where $\mathcal{S}_{ij}(\varphi) = -\int_{\Gamma_j} \varphi_j(y) F(x-y) \, ds_y$, $x \in \Gamma_i$; $i,j = 1,2$;
$\mathcal{S}_{ij} : H^{-1/2}(\Gamma_j) \to H^{1/2}(\Gamma_i)$
and the double layer potential operator $\mathcal{K} : H^{1/2}(\Gamma) \to H^{1/2}(\Gamma)$ with

$$\mathcal{K}(u)(x) = \int_\Gamma u(y) \frac{\partial}{\partial n_y} F(x-y) \, ds_y = \begin{pmatrix} \mathcal{K}_{11} & \mathcal{K}_{12} \\ \mathcal{K}_{21} & \mathcal{K}_{22} \end{pmatrix} \begin{pmatrix} u_1 \\ u_2 \end{pmatrix} \quad (4.27)$$

where $\mathcal{K}_{ij}(u)(x) = \int_{\Gamma_j} u_j(y) \frac{\partial}{\partial n_y} F(x-y) \, ds_y$, $x \in \Gamma_i$; $i,j = 1,2$
$\mathcal{K}_{ij} : H^{1/2}(\Gamma_j) \to H^{1/2}(\Gamma_i)$.

We give a sketch of the doubly connected domain to illustrate the introduced quantities.

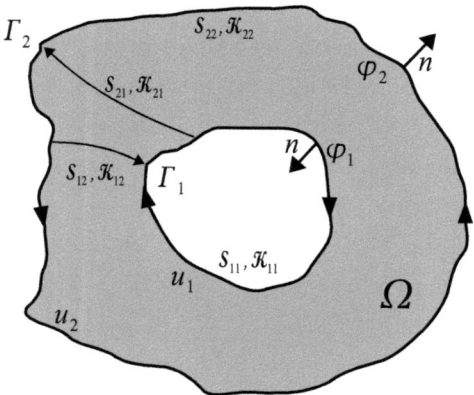

It shows the orientation of the outward normal n and the associated direction of parametrization of Γ_i, $i = 1, 2$.

These definitions and (4.25) provide the following boundary integral equation

$$0 = \frac{u}{2} - \mathcal{K}(u) + \mathcal{S}(\varphi) \quad \text{in} \quad H^{1/2}(\Gamma) \quad (4.28)$$

where $\varphi = \Phi(u) = \frac{1}{\lambda_2} \begin{pmatrix} -q(u_1) \\ \alpha(u_2)(u_2 - u_{env}) \end{pmatrix}$.

Existence and uniqueness of a solution of 4.28

Analogously to 4.1.3 we assume the following conditions

(B1) *Scaling:* $\quad diam(\Omega) < 1$
This implies that $S : H^{-1/2}(\Gamma) \to H^{1/2}(\Gamma)$ is a strongly elliptic operator on the boundary of the multiply connected domain Ω, (see e.g. [45], chap 10.3).

(B2) *Mapping property and strong monotonicity:*
Setting $h = (h_1, h_2)$ and $h_1(s) := -\frac{q(s)}{\lambda_2}$ and $h_2(s) := \frac{\alpha(s)}{\lambda_2}(u - u_{env})$ we require that

$$\frac{h_i(s) - h_i(t)}{s - t} \geq c_i \, ; \quad i = 1,2 \quad \text{and} \quad \min_{1 \leq i \leq 2} c_i =: c_{min} > 0$$

The assumption provides continuity and strong monotonicity of the associated superposition operator $\Phi(u)(x_1, x_2) := h(u_1(x_1), u_2(x_2))$ with $\Phi : H^{1/2}(\Gamma) \to H^{-1/2}(\Gamma)$.

The implication of (B2) is possible since the temperature dependence of the heat flux mapping $h : \mathbb{R}^2 \to \mathbb{R}^2$, $h = (h_i(u_i))_{i=1,2}$ is prescribed in diagonal form; i.e. we have no $h_i(u_j)$ for $i \neq j$. This yields

$$\langle \Phi(u) - \Phi(v), u - v \rangle = \sum_{i=1}^{2} \langle \Phi_i(u_i) - \Phi_i(v_i), u_i - v_i \rangle$$
$$\geq \sum_{i=1}^{2} c_i \|u - v\|^2_{L^2(\Gamma_i)} \geq c_{min} \|u - v\|^2_{L^2(\Gamma)}$$

where $\langle \cdot, \cdot \rangle$ denotes the $\langle \cdot, \cdot \rangle_{H^{-1/2}(\Gamma), H^{1/2}(\Gamma)}$ or $\langle \cdot, \cdot \rangle_{H^{-1/2}(\Gamma_i), H^{-1/2}(\Gamma_i)}$ respectively.

Theorem 4.3
Assume that (B1) and (B2) are satisfied. Then there exists a unique solution $u \in H^{1/2}(\Gamma)$ of (4.28) which is bounded by $\|u\|_{H^{1/2}(\Gamma)} \leq c^2_{emb} \sqrt{\sum_{i=1}^{2} |\Gamma_i| h_i^2(0)}$.

Remark

c_{emb} denotes the constant of the trace embedding between $H^{1/2}(\Gamma)$ and $H^1(\Omega)$ w.r.t. $\|\cdot\|_*$; $\|w\|_*^2 = \|\nabla w\|_{L^2(\Omega)}^2 + c_{min} \|w\|_{L^2(\Gamma)}^2$ denotes the physically consistent norm on $H^1(\Omega)$. Defining $\|u\|_{H^{1/2}(\Gamma)} := \inf\{\|v\|_* : v|_\Gamma = u\}$, we can set $c_{emb} = 1$.

Proof Theorem 4.3

(i) *Existence and Uniqueness*

Assumption (B1) enables us to define the Steklov-Poincaré Operator $\mathcal{P} : H^{1/2}(\Gamma) \to H^{-1/2}(\Gamma)$ on the boundary of a multiply connected domain: Consider equation (4.28) and apply $\mathcal{S}^{-1}$. Noting $\varphi = -\frac{\partial u}{\partial n}$, we define $\mathcal{P} : u \mapsto \frac{\partial u}{\partial n}$ via

$$\mathcal{P} = \mathcal{S}^{-1} \circ \left(\frac{Id}{2} - \mathcal{K}\right).$$

It determines the relation between the Cauchy-Data $\left(u, \frac{\partial u}{\partial n}\right)$ of harmonic functions on the doubly connected domain Ω.

We use the definition of $\mathcal{P}$ and apply $\mathcal{S}^{-1}$ to (4.28). This yields the equivalent equation

$$A(u) := \mathcal{P}(u) + \Phi(u) = 0. \tag{4.29}$$

The hemicontinuity of A is clear. Hence it suffices to show strong monotonicity of A to get existence and uniqueness of a solution of $A(u) = 0$ by the Theorem of Browder and Minty. Using assumption (B2) we get the strong monotonicity of A analogously to the proof of Theorem 4.1.

(ii) *Boundedness*

There holds

$$\langle \Phi(u), u \rangle \geq c_{min} \|u\|_{L^2(\Gamma)}^2 + \langle \Phi(0), u \rangle$$

$$\geq c_{min} \|u\|_{L^2(\Gamma)}^2 - \sqrt{\sum_{i=1}^{2} |\Gamma_i| h_i^2(0)} \, \|u\|_{L^2(\Gamma)}$$

and hence

$$\langle Au, u\rangle \geq \|u\|_*^2 - \sqrt{\sum_{i=1}^{2} |\Gamma_i|\, h_i^2(0)}\, \|u\|_{L^2(\Gamma)}$$

$$\geq \frac{1}{c_{emb}^2} \|u\|_{H^{1/2}(\Gamma)}^2 - \sqrt{\sum_{i=1}^{2} |\Gamma_i|\, h_i^2(0)}\, \|u\|_{H^{1/2}(\Gamma)}$$

which provides the stated bound. $\square$

4.2.5. Iterative determination of the boundary temperatures

Analogously to section 4.1.4 we propose a fixed point iteration based on Banach's fixed point Theorem. The main difference is that the single layer operator $\mathcal{S}$ defined by (4.26) is not self-adjoint on multiply connected domains. W.r.t. the introduced vector notation we consider the operator $\mathcal{T}: H^{1/2}(\Gamma) \to H^{1/2}(\Gamma)$ with $\mathcal{T}(u) := u/2 - \mathcal{K}(u) + \mathcal{S}(\Phi(u))$. Again we consider the fixed point equation $\mathcal{G}_\gamma(u) = u$ defined by (4.9). Its solution exists uniquely due to Theorem 4.3. We determine a γ which ensures that $\mathcal{G}_\gamma$ is a contraction in $H^{1/2}(\Gamma)$. First we verify Lipschitz-continuity of $\mathcal{T}$ using the following inforced assumption on h.

(B2') *Monotonicity and Lipschitz continuity of h*
Using the notation of (B2) We require that h is Lipschitz-continuous satisfying

$$|h_i(s) - h_i(t)| \leq C_i\, |s-t| \quad i=1,2 \quad \text{and} \quad \max_{1\leq i \leq 2} C_i =: C_{max} < \infty$$

such that the respective superposition operator $\Phi(u)(x) := h(u(x))$, $\Phi: H^{1/2}(\Gamma) \to H^{-1/2}(\Gamma)$ is Lipschitz continuous. Moreover h shall satisfy the monotonicity estimate in (B2) such that Φ is strongly monotonous.

Let us give an example for a suitable h that satisfies the condition (B2') and thus (B2) in both components, i.e. for $x \in \Gamma_1$ and $x \in \Gamma_2$. (B2') holds true in the first component with the heat flow density $\tilde{q} = \tilde{q}(u)$ in view of the linear-affine resistivity $\rho(u) := \rho_0 \left(1 + \alpha_\rho (u - u_0)\right)$, $\alpha_\rho > 0$. In the second component (B2') is satisfied e.g. for the truncation and extension of the monotone and continuous heat transfer coefficient α in (3.3). With these settings (B2') is satisfied with

$$c_{min} = \frac{\min(\alpha_l, c_0)}{\lambda_2} \text{ and } C_{max} = \frac{\max(\alpha_h, c_0)}{\lambda_2} \text{ where } c_0 = \frac{\rho_0 \alpha_\rho I^2}{|\Gamma_1| |\Omega_1|}. \quad (4.30)$$

For the strong monotonicity condition of Φ namely $\min(\alpha_l, c_0) > 0$ we require $I > 0$. This is no restriction since $I = 0$ implies $u \equiv u_{env}$.

Lemma 4.4 (Lipschitz continuity of $\mathcal{T}$)
Suppose and (B2'). Then there exists $\tilde{L} > 0$ such that

$$\|\mathcal{T}(u) - \mathcal{T}(v)\|_{H^{1/2}(\Gamma)} \leq \tilde{L} \, \|u - v\|_{H^{1/2}(\Gamma)} \quad \text{for } u, v \in H^{1/2}(\Gamma).$$

The proof follows directly from Lemma 4.2 applied componentwise.

Symmetric bilinear form on $H^{1/2}(\Gamma)$

In order to show strong monotonicity of $\mathcal{T}$ we introduce an equivalent norm on $H^{1/2}(\Gamma)$ induced by the inverse of the single layer operator $\mathcal{S}^{-1}$. Here we vary the approach in section 4.1.3 since the bilinear form $\langle u, \mathcal{S}^{-1}(v)\rangle_{L^2(\Gamma)}$ is not symmetric in the multiply connected domain case. Therefore we introduce an alternative representation of $\mathcal{T}$ using the diagonal components of $\mathcal{S}$. We define

$$\tilde{\mathcal{S}}_{ij}(\varphi) := \begin{cases} \mathcal{S}_{ij}(\varphi) & \text{for } i = j \\ 0 & \text{for } i \neq j \end{cases}, \quad \tilde{\mathcal{K}}_{ij}(u, \varphi) := \begin{cases} \mathcal{K}_{ij}(u) & \text{for } i = j \\ \mathcal{K}_{ij}(u) - \mathcal{S}_{ij}(\varphi) & \text{for } i \neq j \end{cases}$$

and set $\mathcal{S}_d := (\tilde{\mathcal{S}}_{ij})_{1 \leq i,j \leq 2}$ and $\mathcal{K}_d := (\tilde{\mathcal{K}}_{ij})_{1 \leq i,j \leq 2}$. Thus the operator $\mathcal{T} : H^{1/2}(\Gamma) \to H^{1/2}(\Gamma)$ reads as $\mathcal{T}(u) = u/2 - \mathcal{K}_d(u, \Phi(u)) + \mathcal{S}_d(\Phi(u))$. Now

- assuming (B1) - we can introduce the symmetric bilinear form

$$\langle u, v \rangle_{\mathcal{S}_d^{-1}(\Gamma)} := \langle u, \mathcal{S}_d^{-1}(v) \rangle_{L^2(\Gamma)} \; ; \; u, v \in H^{1/2}(\Gamma)$$

and the associated norm $\|u\|_{\mathcal{S}_d^{-1}(\Gamma)}^2 := \langle u, \mathcal{S}_d^{-1}(u) \rangle_{L^2(\Gamma)}$. A componentwise implication from section 4.1.4 yields that it is equivalent to the Sobolev-Slobodetskii-norm on $H^{1/2}(\Gamma)$ i.e.

$$\exists c_\Gamma > 0 : \quad \frac{1}{c_\Gamma} \|u\|_{\mathcal{S}_d^{-1}(\Gamma)} \leq \|u\|_{H^{1/2}(\Gamma)} \leq c_\Gamma \|u\|_{\mathcal{S}_d^{-1}(\Gamma)} \, .$$

Definition of a modified Steklov-Poincaré Operator on doubly connected domains

Consider $\mathcal{T}(u) = u/2 - \mathcal{K}_d(u, \Phi(u)) + \mathcal{S}_d(\Phi(u)) = 0$ and apply $\mathcal{S}_d^{-1}$. This yields the nonlinear map

$$\mathcal{P}_d : u \mapsto \frac{\partial u}{\partial n} \; ; \; \mathcal{P}_d(u) := \mathcal{S}_d^{-1}\left(\frac{u}{2} - \mathcal{K}_d(u, \Phi(u))\right) \; ; \; \Phi(u) = -\frac{\partial u}{\partial n}. \quad (4.31)$$

It determines the relation between the Cauchy-data $(u, \frac{\partial u}{\partial n})$ of harmonic functions in doubly connected domains using the nonlinear superposition operator $\Phi = \Phi(u)$.

Remark
The defintion above is applicable to connected domains with arbitrary multiplicity. Observe that for simply connected domains $\mathcal{P}_d$ coincides with the classical Steklov-Poincaré operator, since we have $\mathcal{S}_d = \mathcal{S}$ and $\mathcal{K}_d = \mathcal{K}$ in this case.

Lemma 4.5 (Strong monotonicity of $\mathcal{T}$)
Suppose (B1) and (B2'). Then there exists $m > 0$ such that

$$\langle \mathcal{T}(u) - \mathcal{T}(v), \mathcal{S}_d^{-1}(u-v) \rangle_{L^2(\Gamma)} \geq m \, \|u-v\|_{\mathcal{S}_d^{-1}(\Gamma)}^2 \quad \textit{for} \quad u, v \in H^{1/2}(\Gamma).$$

Proof
We use the representation $\mathcal{T}(u) = u/2 - \mathcal{K}_d(u, \Phi(u)) + \mathcal{S}_d(\Phi(u))$. $\mathcal{S}_d^{-1}$ is self adjoint, hence we have

$$\langle \mathcal{T}(u) - \mathcal{T}(v), \mathcal{S}_d^{-1}(u-v) \rangle_{L^2(\Gamma)} = \langle Au - Av, u-v \rangle_{L^2(\Gamma)}$$

where $A = \mathcal{P}_d + \Phi$ and $\mathcal{P}_d$ denotes the modified Steklov-Poincaré operator defined in (4.31). Thus there holds

$$\langle Au - Av, u-v \rangle = \langle \mathcal{P}_d(u) - \mathcal{P}_d(v), u-v \rangle + \langle \Phi(u) - \Phi(v), u-v \rangle$$

where we abbreviated $\langle \cdot, \cdot \rangle_{L^2(\Gamma)} = \langle \cdot, \cdot \rangle$. Despite the modification, $\mathcal{P}_d$ maps the Dirichlet data of harmonic functions to the respective Neumann data. Hence, using the divergence Theorem, we get

$$\langle \mathcal{P}_d(u) - \mathcal{P}_d(v), u-v \rangle = \int_\Omega |\nabla w_0|^2 \, dx$$

where $w_0 \in H^1(\Omega)$ denotes the harmonic extension of the Cauchy-data $(u-v) \in H^{1/2}(\Gamma)$ and $(\mathcal{P}_d(u) - \mathcal{P}_d(v)) \in H^{-1/2}(\Gamma)$ to Ω. Using the monotonicity assumption on Φ in (B2') we get $\langle Au - Av, u-v \rangle_{L^2(\Gamma)} \geq \|u-v\|_\star^2$ where $\|w\|_\star^2 = \|\nabla w\|_{L^2(\Omega)}^2 + c_{min} \|w\|_{L^2(\Gamma)}^2$ denotes the physically consistent norm on $H^1(\Omega)$. Finally the estimates

$$\|u-v\|_\star^2 \geq \frac{1}{c_{emb}^2} \|u-v\|_{H^{1/2}(\Gamma)}^2 \geq \frac{1}{c_{emb}^2 c_\Gamma^2} \|u-v\|_{\mathcal{S}_d^{-1}(\Gamma)}^2$$

yield the statement of Lemma 4.5 with $m = \frac{1}{c_{emb}^2 c_\Gamma^2}$. □

Construction of the iterative sequence

We define an iterative sequence $(u^{(n)})_{n \in \mathbb{N}} \subset H^{1/2}(\Gamma)$ which converges to the solution of (4.28) for an arbitrary initial function $u^{(1)} \in H^{1/2}(\Gamma)$. Before doing so we observe that the Lipschitz estimate in Lemma 4.4 also holds w.r.t. the $\mathcal{S}_d^{-1}$-norm on $H^{1/2}(\Gamma)$.

Moreover by (B1), $\mathcal{S}_d : H^{-1/2}(\Gamma) \to H^{1/2}(\Gamma)$ is a strongly elliptic, self-adjoint operator and so is $\mathcal{S}^{-1} : H^{1/2}(\Gamma) \to H^{-1/2}(\Gamma)$. Thus the bilinear form $\langle u, v \rangle_{\mathcal{S}_d^{-1}(\Gamma)}$ is symmetric and we obtain the following result.

Theorem 4.4
Let the assumptions (B1) and (B2') hold. Define the iterative sequence $(u^{(n)})_{n \in \mathbb{N}} \subset H^{1/2}(\Gamma)$ by $u^{(n+1)} := \mathcal{G}_\gamma(u^{(n)})$, $\gamma = m/L^2$ where $L = c_\Gamma \tilde{L}$ denotes the $\mathcal{S}_d^{-1}$- Lipschitz constant and m the $\mathcal{S}_d^{-1}$-monotonicity constant of $\mathcal{T}$. Then, for every initial function $u^{(1)} \in H^{1/2}(\Gamma)$, $(u^{(n)})_{n \in \mathbb{N}}$ converges to the solution u of (4.28) with respect to $\|\cdot\|_{\mathcal{S}^{-1}}$ with the a priori error estimate

$$\|u^{(n)} - u\|_{\mathcal{S}_d^{-1}(\Gamma)} \leq \frac{k^n}{1-k} \|u^{(2)} - u^{(1)}\|_{\mathcal{S}_d^{-1}(\Gamma)} \ , \quad k = \sqrt{1 - \frac{m^2}{L^2}} \ .$$

The proof is analogous to the proof of Theorem 4.2.

4.2.6. The case of a multiply connected domain

In this section we extend our previous considerations from a doubly connected domain to a multiply connected one. Hence we can treat electrical cables with an ensemble of conductors with possibly different current loads.

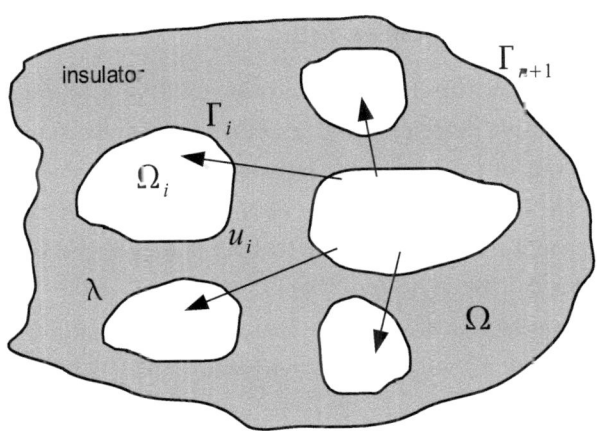

We use the following notation: N denotes the quantity of conductor cross sections, Ω_i are the conductor cross sections $i = 1, \ldots, N$, $\Gamma_i = \partial \Omega_i$ its boudaries, Ω is the insulator cross section, Γ_{N+1} denotes the (outer-) insulator boundary, $u_j \in H^{1/2}(\Gamma_j)$ denote the boundary temperatures $j = 1, \ldots, N+1$, $q_i = q(u_i)$ is the heat flux over Γ_i, λ denotes the heat conductivity of the insulator.

For $\Gamma = \partial \Omega = \bigcup_{j=1}^{N+1} \Gamma_j$ the corresponding function spaces $H^s(\Gamma)$, $s \in \{-\frac{1}{2}, \frac{1}{2}\}$ are given by $H^s(\Gamma) = \prod_{j=1}^{N+1} H^s(\Gamma_j)$ and $\|\cdot\|_{H^s(\Gamma)}^2 = \sum_{j=1}^{N+1} \|\cdot\|_{H^s(\Gamma_j)}^2$.

The boundary value problem

$$-\Delta u = 0 \text{ in } \Omega; \quad \lambda \frac{\partial u}{\partial n} = q_i(u) \text{ on } \Gamma_i; \quad -\lambda \frac{\partial u}{\partial n} = \alpha(u)(u - u_{env}) \text{ on } \Gamma_{N+1}$$

leads to a boundary integral equation for
$u = (u_1, \ldots, u_{N+1}) \in H^{1/2}(\Gamma)$: $\frac{u}{2} - \mathcal{K}(u, \Phi(u)) + \mathcal{S}(\Phi(u)) = 0$.
With $\varphi = \Phi(u) = \frac{1}{\lambda}(-q_1(u_1), \ldots, -q_N(u_N), \alpha(u_{N+1})(u_{N+1} - u_{env}))$. The single and double layer potential operators $\mathcal{S}$ and $\mathcal{K}$ are defined in the same way as in (4.26), (4.27) for $i, j = 1, \ldots, N+1$. With these settings Theorem 4.3 applies to the multiply connected domain case.

Application to multiwire cables

Now we will see how the crucial assumption (B2') of Theorem 4.4 is satisfied and how the iterative determination is realized in applications.
If the material out of the conductor cross sections is inhomogeneous (e.g. air gaps between the insulator material), then the constant heat conductivity λ of the insulator material, can be replaced by a homogenized heat conductivity $\bar{\lambda}$. Here we refer to [26], [47], [55], [74].
The estimate from section 4.2.1 can be applied for each conductor cross section separately. Thus again, we use the approximate heat flow densities over the

boundary of the conductor cross section for $i = 1, \ldots, N$ as

$$q_i = q_i(u_i) = \frac{\rho_i(u_i) I_i^2}{|\Gamma_i| |\Omega_i|}$$

with $\rho_i(u_i) = (\rho_0)_i (1 + (\alpha_\rho)_i (u_i - u_0))$. The indexed quantities have the same meaning as before.

Moreover we use the truncated heat transfer coefficient α from (3.3). Thus the associated boundary mappings $h_j : \mathbb{R} \to \mathbb{R}$; $j = 1, \ldots, N+1$ with

$$h_i(u_i) := \frac{q_i(u_i)}{\lambda}; \; i = 1, \ldots, N \quad \text{and} \quad h_{N+1}(u_{N+1}) := \frac{\alpha(u_{N+1})}{\lambda}(u_{N+1} - u_{env})$$

fulfill the assumption (B2') with the following bounds

$$c_{min} = \frac{\min(\alpha_l, b_{min})}{\lambda} \quad \text{and} \quad c_{max} = \frac{\max(\alpha_h, b_{max})}{\lambda} \quad (4.32)$$

where $b_{min} = \min\limits_{1 \le i \le N} \left\{ \frac{(\rho_0)_i (\alpha_\rho)_i I_i^2}{|\Gamma_i| |\Omega_i|} \right\}$ and $b_{max} = \max\limits_{1 \le i \le N} \left\{ \frac{(\rho_0)_i (\alpha_\rho)_i I_i^2}{|\Gamma_i| |\Omega_i|} \right\}$.

Hence, for the strong monotonicity of $\mathcal{T}$, we need the restricitve assumption $I_i > 0$; $i = 1, \ldots, N$. It is possible to elude this assumption considering cross sections with $I_i > 0$ only; the currentless cross sections are included in the insulator domain Ω and can be taken into account when the homogenized heat conductivity $\overline{\lambda}$ is computed. Since this approach is cumbersome w.r.t. possibly changing current loads of the cable, we propose an alternative where the monotonicity of $\mathcal{T}$ does not depend on the current I_i; $i = 1, \ldots, N$. It uses a property of the Cauchy data of harmonic functions in certain multiply connected domains.

Damping property

Let $u = (u_1, \ldots, u_{N+1}) \in H^{1/2}(\Gamma)$ and $\varphi = (\varphi_1, \ldots, \varphi_{N+1}) \in H^{-1/2}(\Gamma)$ denote a solution of $0 = \left(\frac{Id}{2} - \mathcal{K}\right)(u) + \mathcal{S}(\varphi)$. Consider the linear Steklov-Poincaré Operator $\mathcal{P} : u \mapsto \frac{\partial u}{\partial n}$ defined by $\mathcal{P} = \mathcal{S}^{-1} \circ \left(\frac{Id}{2} - \mathcal{K}\right)$. Now we extract its diagonal components $\mathcal{P}_{jj} : H^{1/2}(\Gamma_j) \to H^{-1/2}(\Gamma_j)$; $j = 1, \ldots, N+1$ defined by the matix valued notation of $\mathcal{S}$ and $\mathcal{K}$ in (4.26), (4.27).

Definition 4.1 (The damping property)
Γ has the damping property if

$$\min_{1\leq i\leq N} m_i \geq m_{N+1} \quad \text{where} \quad m_j = \inf_{v\in H^{1/2}(\Gamma_j)\setminus\{0\}} \frac{\|\mathcal{P}_{jj}(v)\|_{H^{-1/2}(\Gamma_j)}}{\|v\|_{\mathcal{S}_d^{-1}{}_{jj}(\Gamma_j)}}. \quad (4.33)$$

This property means that a change of the boundary temperature changes the inner normal derivatives more than the outer normal derivative.

For domains with the damping property the lower bounds in (4.30) and (4.32) read as $c_{min} = \frac{\alpha_l}{\lambda}$. Moreover, if (4.33) is verified by an a priori estimate, there is no need to exclude the case $I = 0$. Now Theorems 4.3 and 4.4 can be applied analogously to the doubly connected domain case.

Remark

In the preceding Definition we consider a diagonalized, i.e. reduced form of a damping property which possibly can be formulated more generally; respecting also the nondiagonal influence of the boundary temperatures on its normal derivatives. Such a generalization is not necessary in our applicational context, since the heat flux $h = (h_i(u_i))_{i=1,\ldots,N+1}$ is prescribed in diagonal form, i.e. we have no $h_i(u_j)$ for $i \neq j$, as noticed in assumption (B2).

4.2.7. The case of rotational symmetry

Finally we treat the outer domain boundary value problem (4.18), (4.19) with a rotationally symmetric cross section. This case can be used as a benchmark example for the iteration in Theorem 4.4 or for boundary element methods solving (4.28). We use the notation of section 3.2.3 where r_1 is the inner radius of the insulator, r_2 denotes the outer radius of the insulator, u_1 is the inner boundary temperature at $\Gamma_1 = \partial B_{r_1}$ and u_2 denotes the outer boundary temperature at $\Gamma_2 = \partial B_{r_2}$.

Without loss of generality we can choose a suitable unit for the radius such that the relation $0 < r_1 < r_2 < 1/2$ - and thus assumption (B1) - is fulfilled. Due to the rotational symmetry of the system, the boundary temperatures u_1 and u_2 are constant.

Boundary integral operators on the boundary of an annulus

Now we specify $\mathcal{T} = \frac{Id}{2} - \mathcal{K} + \mathcal{S} \circ \Phi$ for constant boundary temperatures $u = (u_1, u_2)^T \in H^{1/2}(\Gamma) \cap \mathbb{R}^2$ and for constant heat flux $\varphi = (\varphi_1, \varphi_2)^T \in H^{-1/2}(\Gamma) \cap \mathbb{R}^2$. There holds

$$\mathcal{K} = \frac{1}{2}\begin{pmatrix} 1 & 1 \\ 1 & 1 \end{pmatrix} \quad \text{and thus} \quad \frac{Id}{2} - \mathcal{K} = -\frac{1}{2}\begin{pmatrix} 0 & 1 \\ 1 & 0 \end{pmatrix}$$

$$\text{and} \quad \mathcal{S} = -\begin{pmatrix} r_1 \ln r_1 & r_2 \ln r_2 \\ r_1 \ln r_2 & r_2 \ln r_2 \end{pmatrix} \quad ; \quad \mathcal{S}_d^{-1} = -\begin{pmatrix} \frac{1}{r_1 \ln r_1} & 0 \\ 0 & \frac{1}{r_2 \ln r_2} \end{pmatrix}.$$

Observe that $\mathcal{S}$ is not symmetric. The eigenvalues are

$$\lambda_{1,2} = -\frac{r_1 \ln r_1 + r_2 \ln r_2}{2} \pm \sqrt{\left(\frac{r_1 \ln r_1 - r_2 \ln r_2}{2}\right)^2 - r_1 r_2 (\ln r_2)^2} > 0$$

for $0 < r_1 < r_2 < 1/2$; i.e. $\mathcal{S}$ is positive definite and thus invertible. Using the identifications of the boundary integral operators, $\mathcal{T}(u) = 0$ reads as

$$u_2/2 = r_1 \ln r_1 \varphi_1 + r_2 \ln r_2 \varphi_2 \tag{4.34}$$
$$u_1/2 = r_1 \ln r_2 \varphi_1 + r_2 \ln r_2 \varphi_2$$

where $\begin{pmatrix} \varphi_1 \\ \varphi_2 \end{pmatrix} = \begin{pmatrix} h_1(u_1) \\ h_2(u_2) \end{pmatrix} = \frac{1}{\lambda_2}\begin{pmatrix} -q(u_1) \\ \alpha(u_2)(u_2 - u_{env}) \end{pmatrix}.$

Screening effect of the single layer operator

Using the matrix valued definition of $\mathcal{S}$ in (4.26), we observe that $\mathcal{S}_{12} = r_2 \ln r_2 = \mathcal{S}_{22}$. For multiply connected domains we have in general

$$\mathcal{S}_{1,N+1} = \ldots = \mathcal{S}_{N,N+1} = \mathcal{S}_{N+1,N+1}.$$

This is the screening effect of the outer boundary Γ_{N+1} for the single layer operator.

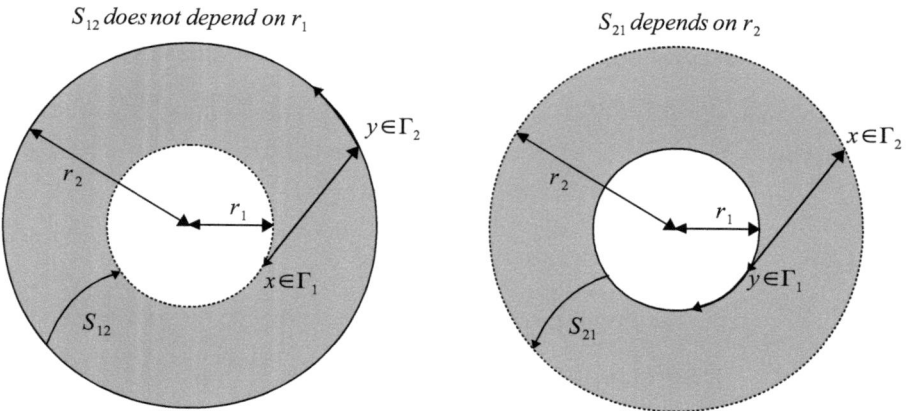

The outer boundary Γ_2 has the screening effect since the integration over Γ_2 'does not see' the position of any point x in the interior of Γ_2 and in particular not the position of $x \in \Gamma_1$; hence $\mathcal{S}_{12}$ does not depend on r_1. On the other hand an integration over Γ_1 does not compensate the distance to the outer boundary Γ_2; thus $\mathcal{S}_{21}$ depends on r_2

This effect has its physical counterpart when considering gravitational or electrical fields. The gravitational or electrostatical potential on Γ_2 does not depend on the position of the mass particle / electron in the interior of Γ_2.

Verification of the damping property

Proposition 4.4
Suppose that u satisfies (4.28) i.e. $\mathcal{T}(u) = 0$ specified as above. Then

$$\left| \frac{\partial}{\partial u_1} \varphi_1 \right| = \frac{r_2}{r_1} \left| \frac{\partial}{\partial u_2} \varphi_2 \right| \geq \left| \frac{\partial}{\partial u_2} \varphi_2 \right| \geq \frac{\alpha_l}{\lambda_2}. \qquad (4.35)$$

Proof
$\mathcal{S}$ is invertible and the Steklov-Poincaré operator reads as

$$\mathcal{P} = \mathcal{S}^{-1} \circ \left(\frac{Id}{2} - \mathcal{K}\right) = \frac{1}{2\ln(r_1/r_2)} \begin{pmatrix} \frac{-1}{r_1} & \frac{1}{r_1} \\ \frac{\ln r_1}{r_2 \ln r_2} & \frac{-1}{r_2} \end{pmatrix}$$

Hence its diagonal components are given by $\mathcal{P}_{jj} = -\frac{1}{2r_j \ln(r_1/r_2)}$. Thus, with the notation of (4.33) we have $m_j = C \left|\frac{u_j}{2r_j \ln(r_1/r_2) u_j}\right|$; $j = 1, 2$ where $C = \frac{\|1\|_{H^{-1/2}(\Gamma_j)}}{\|1\|_{\mathcal{S}_d^{-1}(\Gamma_j)}}$ and $\|1\|_{H^{-1/2}(\Gamma_j)}$ is chosen such that the quotient C does not depend on j. This implies $m_2 = \frac{r_1}{r_2} m_1$. In the case of rotational symmetry we have $m_j = \frac{\partial \varphi_j}{\partial u_j}$. An alternative derivation of the equality in (4.35) is the differentiation of (4.34) w.r.t. u_1 and u_2. Then, using $r_2 > r_1$, the outer boundary condition $\varphi_2 = \frac{\alpha(u_2)}{\lambda_2}(u_2 - u_{env})$ and the truncation of α in (3.3), yields the statement of Proposition 4.4. □

With the estimates for Φ in (B2') and the damping property we obtain the Lipschitz and the monotonicity constants of $\mathcal{T}$. This is essential for the error estimate in the iterative scheme of Theorem 4.4. We use $\|\cdot\|^2_{\mathcal{S}_d^{-1}(\Gamma)} := \|\cdot\|^2_{\mathcal{S}_d^{-1}(\partial B_{r_1})} + \|\cdot\|^2_{\mathcal{S}_d^{-1}(\partial B_{r_2})}$.

Lipschitz estimate
As $\|1\|_{\mathcal{S}_d^{-1}(\partial B_r)} = \sqrt{\frac{-2\pi}{\ln r}}$, $r < 1/2$, we have
$\|u - v\|^2_{\mathcal{S}_d^{-1}(\Gamma)} = -2\pi \sum_{i=1}^{2} \frac{(u_i - v_i)^2}{\ln r_i} \geq \frac{-2\pi}{\ln r_1}|u - v|^2$. On the other hand, the Lipschitz continuity of $\bar{\varrho}$ yields for $C_{max} = \frac{1}{\lambda_2} \max(\alpha_h, c_0)$:

$$\|\mathcal{T}(u) - \mathcal{T}(v)\|^2_{\mathcal{S}_d^{-1}(\Gamma)} = \langle \mathcal{T}(u) - \mathcal{T}(v), \mathcal{S}_d^{-1}(\mathcal{T}(u) - \mathcal{T}(v))\rangle_{L^2(\Gamma)}$$
$$\leq -2\pi (u-v)^T B^T \underbrace{\begin{pmatrix} \frac{1}{\ln r_1} & 0 \\ 0 & \frac{1}{\ln r_2} \end{pmatrix}}_{=:-A_L} B(u-v)$$

Where $B = \frac{Id}{2} - \mathcal{K} + C_{max} \mathcal{S}$. Let λ_{max} denote the maximal eigenvalue of A_L, then there holds $\|\mathcal{T}(u) - \mathcal{T}(v)\|^2_{\mathcal{S}_d^{-1}(\Gamma)} \leq 2\pi \lambda_{max} |u - v|^2$. Thus we obtain the Lipschitz constant $L = \sqrt{-\lambda_{max} \ln r_1}$.

Monotonicity estimate

With the damping property the monotonicity of Φ yields for $c_{min} = \frac{\alpha_L}{\lambda_2}$:

$$\langle \mathcal{T}(u) - \mathcal{T}(v), \mathcal{S}_d^{-1}(u-v) \rangle_{L^2(\Gamma)} \geq -2\pi (u-v)^T B^T \underbrace{\begin{pmatrix} \frac{1}{\ln r_1} & 0 \\ 0 & \frac{1}{\ln r_2} \end{pmatrix}}_{=: -A_m} (u-v).$$

where $B = \left(\frac{Id}{2} - \mathcal{K} + c_{min}\mathcal{S}\right)$. A_m is positive definite for every $c_{min} > 0$ and $0 < r_1 < r_2 < 1/2$. Let λ_{min} denote the minimal eigenvalue of the symmetric part of A_m then $\langle \mathcal{T}(u) - \mathcal{T}(v), \mathcal{S}_d^{-1}(u-v) \rangle_{L^2(\Gamma)} \geq 2\pi\lambda_{min}|u-v|^2$. Analogously we get $\|u-v\|_{\mathcal{S}_d^{-1}(\Gamma)}^2 = -2\pi \sum_{i=1}^2 \frac{(u_i-v_i)^2}{\ln r_i} \leq \frac{-2\pi}{\ln r_2}|u-v|^2$. Hence we arrive at the monotonicity constant $m = -\lambda_{min}\ln r_2$.

4.2.8. An application to physical data

We fix some physical data with

Temperatures: $u_0 = 20$, $u_{env} = 50$

Conductor parameters: $\lambda_1 = 400$, $\rho_0 = 1.72 * 10^{-8}$, $\alpha_\rho = 3.83 * 10^{-3}$, $r_1 = 7 * 10^{-4}$

Insulator parameters: $\lambda_2 = 0.17$, $\epsilon = 0.93$, $r_2 = 1 * 10^{-3}$.

Considering the case $I \leq 30$, we obtain the $\mathcal{S}_d^{-1}$-Lipschitz- and the $\mathcal{S}_d^{-1}$-monotonicity constant of $\mathcal{T}$ with $L = 1,71$ and $m = 0,34$. Thus the fixed point mapping $\mathcal{G}_\gamma$ of Theorem 4.4 is given by $\gamma := \frac{m}{L^2} = 0.117$ and is contractive with $k = 0.9797$. For $u^{(1)} \equiv u_{env}$ the a priori error estimate of the corresponding iteration reads for $n \geq 800$ as

$$\|u^{(n)} - u\|_{\mathcal{S}_d^{-1}(\Gamma)} \leq \frac{k^n}{1-k} \|\mathcal{G}_\gamma(u_{env}) - u_{env}\|_{\mathcal{S}_d^{-1}(\Gamma)}$$

$$\leq \frac{k^n}{1-k} \sqrt{2}\gamma \|1\|_{\mathcal{S}_d^{-1}(\partial B_{r_2})} \sqrt{\left|-u_{env}/2 + r_1 \ln r_1 \frac{q(u_{env})}{\lambda_2}\right|}$$

$$\leq 5,9 * 10^{-6}.$$

We iterate the sequence $\mathcal{G}_\gamma(u^{(n)})$ which is shown in the following figure.

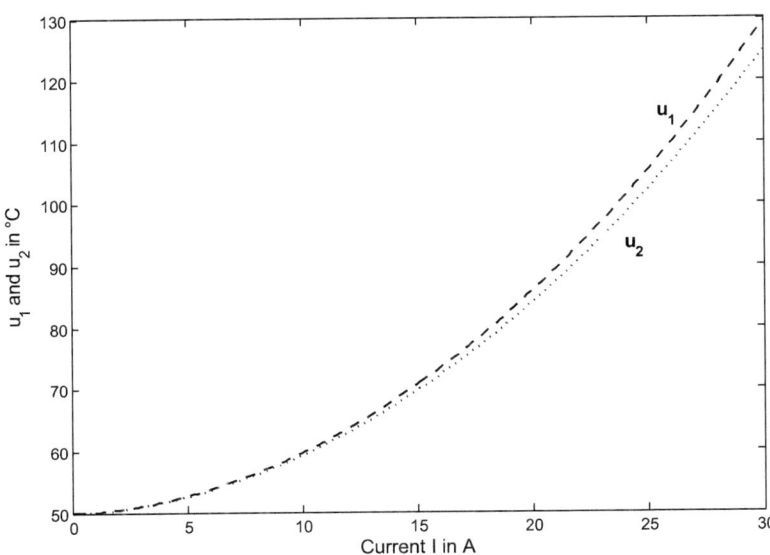

We obtain a very good agreement between these calculated temperatures and experimental results.

5. Conclusions

In chapter 2 and 3 we demonstrated explicitly the connection between the semilinear parabolic equation (2.1) and the cross-sectional boundary value problem (2.22). Provided the data in (2.1) fulfill the subresonance condition of Theorem 2.1, we were able to give explicit asymptotic estimates between the solutions of the stepwise reduced problem. In chapter 4 we proposed an iterative method solving the nonlinear boundary integral equation which is equivalent to (2.22). The presented work also led to some new problems whose treatment shall be of further interest.

The accuracy of our asymptotic estimates depends essentially on the estimate of Friedrichs constant $c_\star$. As mentioned in section 2.4.2, the estimates given in Proposition 2.6 are rather rough. Nevertheless, they give a basic dependence of $c_\star$ w.r.t. geometrical and physical parameters of the domain Ω. Hence it will be valuable to obtain optimal estimates of $c_\star$ with respect to a - possibly different - physically consistent norm $\|\cdot\|_\star$ on $H^1(\Omega)$. Moreover even the optimal estimates can be sharpened in specific cases, using a priori known properties of the solution u of (2.1) or (2.22) respectively.

In the asymptotic analysis part as well as in the treatment by boundary integral equations, we used the case of rotational symmetry as a benchmark example illustrating our results. It was mentioned that this special case is a plausible idealization, since electric mains are rotationally symmetric in many cases. Now we want to sketch how solutions in a rotationally symmetric domain $\Omega \subset \mathbb{R}^d$ change, if the domain is perturbed by a transforming velocity field $V \in C^1(\mathbb{R}^d, \mathbb{R}^d)$. To this end we use the perturbation of identity $\Omega_\epsilon := \{x + \epsilon V(x), \, x \in \Omega\} \, ; \, \epsilon \geq 0$ introduced in the fundamental paper for optimal design in [63]. Consider $\Omega = B_r(0) \subset \mathbb{R}^d$ and e.g. the stationary cross-sectional problem

$$-\lambda \Delta u = \frac{\rho_0 (1 + \alpha_\rho u) I^2}{|B_r|^2} \quad \text{in } B_r(0)$$

$$-\lambda \frac{\partial u}{\partial n} = \alpha (u - u_{env}) \quad \text{on } \partial B_r(0).$$

On the other hand assume that u_ϵ is the solution of the perturbed problem

$$-\lambda \Delta u_\epsilon = \frac{\rho_0 (1 + \alpha_\rho u_\epsilon) I^2}{|\Omega_\epsilon|^2} \quad \text{in } \Omega_\epsilon$$

$$-\lambda \frac{\partial u}{\partial n} = \alpha (u_\epsilon - u_{env}) \quad \text{on } \partial \Omega_\epsilon.$$

Define now a functional $J_\epsilon : H^1(\Omega_\epsilon) \to \mathbb{R}$, $\epsilon > 0$ which assigns a characteristic value of u_ϵ, e.g. $J_\epsilon(u_\epsilon) = \|u_\epsilon\|_{L^\infty(\Omega_\epsilon)}$. Then $|J_\epsilon(u) - J_0(u))|$ describes the shape sensitivity of the functional J with respect to the transforming velocity field V. We remark that this shape sensitivity is the numerator of the directional shape derivative $dJ(\Omega; V) = \lim_{\epsilon \to 0} \frac{J_\epsilon(u_\epsilon) - J_0(u)}{\epsilon}$ introduced in [63]. An estimate $|J_\epsilon(u_\epsilon) - J_0(u)| \leq f(\epsilon)$, $f(\epsilon) \xrightarrow[\epsilon \to 0]{} 0$ will provide information how the solution u is perturbed, if the rotationally symmetric shape Ω is perturbed by ϵV. For an investigation of shape transformation and optimal shapes in the context of heat transfer in electrical cables we refer to [53].

In chapter 4 we introduced the damping property (4.33) as a geometrical property of boundaries of certain domains. This property can be seen as a natural property of the insulator, i.e. of harmonic functions w.r.t. the considered boundary conditions. We verified it in the case of rotational symmetry. It is an outstanding problem to identify a broader class of domains which satisfy the damping property. It is essential for obtaining a numerically acceptable monotonicity constant in Theorem 4.4, especially for low currents. This motivates a study of this property for more general situations, non-symmetric domains in particular.

A. Appendix

Theorem A.1 (Banach's fixed point theorem)
Let (X, d) be a complete metric space with a contraction mapping $T : X \to X$, i.e. there exists $0 \leq q < 1$ such that $d(Tx, Ty) \leq q\, d(x, y)$ for all $x, y \in X$. Define the iterative sequence $(x_n)_{n \in \mathbb{N}} \subset X$ by $x_{n+1} = Tx_n$.
Then, for every initial value $x_1 \in X$, the iterative sequence $(x_n)_{n \in \mathbb{N}}$ converges to the unique solution $x^* \in X$ of the fixed point equation $T(x^*) = x^*$ with the following rate of convergence

$$d(x_n, x) \leq \frac{q^n}{1-q} d(x_2, x_1).$$

Proof. cf. ([79], p. 15)

Theorem A.2 (Browder-Minty theorem)
Let X be a real, reflexive Banach space and let $T : X \to X^*$ be bounded, hemicontinuous, coercive and monotone. Then, for every $g \in X^*$ there exists a solution u of the equation

$$T(u) = g.$$

If, in addition, T is strictly monotone, the solution u is unique.

Proof. cf. ([80], p. 556)

Theorem A.3 (Cavalieri's principle)
Let λ^k, $k \in \mathbb{N}$ denote the k-dimensional Lebesgue-measure on $\mathbb{R}^k$ and let $A \subset \mathbb{R}^{p+1}$, $p \in \mathbb{N}$ be Lebesgue-measurable. For $s \in \mathbb{R}$, define the section $A_s = \{x \in \mathbb{R}^p; (x,s) \in A\}$. Then there holds

$$\lambda^{p+1}(A) = \int_\mathbb{R} \lambda^p(A_s) \, ds.$$

Proof. cf. ([31], Prop. 6.24)

Theorem A.4 (Gronwall's inequality)
Let $\beta, u \in C([a,b]) \cap C^1((a,b))$ and let u satisfy the differential inequality

$$u'(t) \leq \beta(t) \, u(t), \quad t \in (a,b).$$

Then u is bounded by the solution of the corresponding differential equation $u' = \beta u$:

$$u(t) \leq u(a) \, \exp\left(\int_a^t \beta(s) \, ds\right), \quad t \in [a,b].$$

Proof. cf. ([76], p. 42)

Theorem A.5 (Jensen's inequality)
Let μ be a positive measure on a σ-algebra $\mathcal{A}$ in a set Ω. If f is a real function in $L^1(\Omega)$, if $a < f(x) < b$ for all $x \in \Omega$, and if φ is convex on (a,b), then

$$\varphi\left(\frac{1}{\mu(\Omega)} \int_\Omega f \, d\mu\right) \leq \frac{1}{\mu(\Omega)} \int_\Omega (\varphi \circ f) \, d\mu.$$

Proof. cf. ([71], p. 62)

Bibliography

[1] C. Allouch, P. Sablonnière, D. Sbibih, M. Tahrichi,
"Product Integration Methods Based on Discrete Spline Quasi-Interpolants and Application to Weakly Singular Integral Equations", *J. Comput. Appl. Math.*, Vol. 233, No. 11, p. 2855-2866, 2010

[2] P. Amster, M. C. Mariani, O. Méndez
"Nonlinear Boundary Conditions for Elliptic Equations", *J. Diff. Eqns.*, No. 144, p. 1-8, 2005

[3] J. Appell, P. Zabreiko,
Nonlinear Superposition Operators, Cambridge University Press, 1990

[4] H. Attouch, G. Buttazzo, G. Michaille
"Variational Analysis in Sobolev and BV Spaces", *MPS-SIAM series on Optimization*, 2005

[5] H. M. Badr,
"Laminar Combined Convection from a Horizontal Cylinder - Parallel and Contra Flow Regimes", *Int. J. Heat Mass Transfer*, Vol. 27, p. 15-27, 1984

[6] G. R. Baker, M. J. Shelley,
"Boundary Integral Techniques for Multi-connected Domains" *J. Comp. Phys.*, Vol. 64, p. 112-132, 1984

[7] V. Barbu,
Nonlinear Semigroups and Differential Equations in Banach Spaces
Kluwer Academic Publishers, 1976

[8] J. Beck, B. Blackwell, S. Clair
Inverse Heat Conduction, J. Wiley, 1985

[9] J. Bouchala,
"Resonance Problems for p-Laplacian", *Mathematics and Computers in Simulation*, Vol.61, 2003, 599–604.

[10] D. Braess,
Finite Elements, Cambridge University Press, 2007

[11] H. Brézis, M. Sibony,
"Méthodes d' Approximation et d' Itération pour les Operateurs Monotones", *Arch. Rational Mech. Anal.*, Vol. 28, No. 1, p. 59-82, 1968

[12] H. Brézis,
Operateurs Maximaux Monotones, North-Holland, 1973

[13] F. E. Browder, W. V. Petryshyn,
"Construction of Fixed Points of Nonlinear Mappings in Hilbert Space", *J. Math. Anal. Appl.*, Vol. 20, p. 197-228, 1967

[14] L. C. Burmeister,
Convective Heat Transfer, J. Wiley, 1993

[15] T. Chiang, J. Kay,
"On Laminar Free Convection from a Horizontal Cylinder", *Proc. of 4th National Congress of Appl. Mech.*, p. 1213-1219, 1962

[16] M. Chipot,
"Asymptotic Analysis for Problems in Large Cylinders"
Elliptic Equations: an Introductory Course, Birkhäuser, p.73-91, 2009.

[17] S. W. Churchill, H. H. S. Chu,
"Correlating Equations for Laminar and Turbulent Free Convection from a Horizontal Cylinder", *Int. J. Heat Mass Transfer*, Vol. 18, p. 1049-1053, 1975

[18] S. W. Churchill, R. Usagi,
"A General Expression for the Correlation of Rates of Transfer and other Phenomena", *American Institute of Chemical Engineers J.*, Vol. 18, p. 1121-1128, 1972

[19] R. Čiegis, A. Ilgevičius, H.-D. Liess, M. Meilūnas, O. Suboč,
"Numerical Simulation of the Heat Conduction in Electrical Cables", *Math. Model. Anal.*, Vol. 12, p. 425-439, 2007

[20] E. A. Coddington, N. Levinson,
Theory of Ordinary Differential Equations, Mc Graw-Hill Ed. , 1984

[21] G. Criscuolo, G. Mastroianni, G. Monegato,
"Convergence Properties of a Class of Product Formulas for Weakly Singular Integral Equations", *Math. Comput.*, Vol. 55, No 191, p. 213-230, 1990

[22] K. Diethelm,
"Uniform Convergence of Optimal Order Quadrature Rules for Cauchy Principal Value Integrals", *J. Comput. Appl. Math.*, Vol 56 , p. 321-329, 1994

[23] P. Drábek and S. B. Robinson,
"Resonance Problems for the p-Laplacian", *Journal of Functional Analysis*, 169, p. 189-200, 1999

[24] P. Drábek,
"The p-Laplacian - Mascot of Nonlinear Analysis", *Acta Math. Univ. Commenianae*, 76, p.85-98, 2007

[25] K. Dvorsky, J. Gwinner, H.-D. Liess,
"A Fixed Point Approach to Stationary Heat Transfer in Electric Cables", *Mathematical Modelling and Analysis*, Vol. 16, No. 2, p. 286-303, 2011

[26] K. Dvorsky, H.-D. Liess, F. Loos
"Two Approaches for Heat Transfer Simulation of Current Carrying Multicables", *accepted in Mathematics and Computers in Simulation*, 2012

[27] M. Efendiev, F. Hamel
"Asymptotic Behavior of Solutions of Semilinear Elliptic Equations in Unbounded Domains: Two approaches", *Advances in Mathematics*, 221 , p. 1237–1261, 2011

[28] Z. Fang, D. Xie, N. Diao, J. Grace, C.J. Lim
"A New Method for solving the Inverse Conduction Problem in Steady Heat Flux Measurement", *Int. J. Heat Mass Transfer*, Vol. 40, p. 3947-3953, 1997

[29] M. Feistauer, K. Najzar, K. Švadlenka
"On a Parabolic Problem with Nonlinear Newton Boundary Conditions", *Comment. Math. Univ. Carolin.*, Vol. 43,p. 429-455, 2002

[30] N. Filonov,
"On an Inequality between Dirichlet and Neumann Eigenvalues for the Laplace Operator", *St. Petersburg Math. J.*, Vol. 16, No. 2, p. 413-416, 2005

[31] G. B. Folland,
Real Analysis: Modern Techniques and Their Applications, Wiley, 1999

[32] A. Frackowiak, N. D. Botkin, M. Cialkowski, K.-H. Hoffmann,
"A Fitting Algorithm for Solving Inverse Problems of Heat Conduction",
Int. J. of Heat Mass Transfer, Vol. 53, p. 2123-2127, 2010

[33] L. Friedlander,
"Some Inequalities between Dirichlet and Neumann Eigenvalues", *Arch. Rational Mech. Anal.*, Vol. 116, p. 153-160, 1991

[34] M. Ganesh, O. Steinbach,
"Nonlinear Boundary Integral Equations for Harmonic Problems",
J. Int. Eqs. Appl., Vol. 11, p. 437-459, 1999

[35] M. Ganesh, O. Steinbach,
"Boundary Element Methods for Potential Problems with Nonlinear Boundary Conditions", *Mathematics of Computation*, Vol. 70, p. 1031-1042, 2001

[36] G. N. Gatica, N. Heuer
"On the Numerical Analysis of Nonlinear Twofold Saddle Point Problems", *IMA, Journal of Numerical Analysis*, Vol. 23, p. 301-330, 2003

[37] G. N. Gatica, M. Maischak, E. P. Stephan
"Numerical Analysis of a Transmission Problem with Signorini Contact Using Mixed FEM and BEM", *ESAIM, Math. Model. Numer. Anal.*, Vol. 45, p. 779-802, 2011

[38] D. Greenspan,
Numerical Solution of Ordinary Differential Equations, J. Wiley, 2005

[39] I. M. Gelfand, N. J. Vilenkin,
Generalized Functions, vol. 4, Academic Press, New York, 1964

[40] D. Gilbarg, N. S. Trudinger,
Elliptic Partial Differential Equations of Second Order, Springer, New York, 2001

[41] A. Greenbaum, L. Greengard, G. B. McFadden,
"Laplace's Equation and the Dirichlet-Neumann Map in Multiply Connected Domains", *J. Comp. Phys.*, Vol. 105, p. 267-278, 1993

[42] N. M. Günter
Potential Theory and its Application to Basic Problems of Mathematical Physics, Gestekhizdat, Moscow, 1953

[43] J. Harada, M. Ôtani,
"H^2-Solutions for some Elliptic Equations with Nonlinear Boundary Conditions", *Discrete and Continuous Dynamical Systems*, p. 333-339, 2009

[44] G. Hsiao , W. Wendland,
"A Finite Element Method for Some Integral Equations of the First Kind", *JMAA*, Vol. 58, p. 449-481, 1977

[45] G. Hsiao , W. Wendland,
Boundary Integral Equations, Springer, 2008

[46] J. G. Kaufman et al.,
Electrical and Magnetic Properties of Metals, ASM International, 2000.

[47] J. B. Keller,
"Effective Conductivity, Dielectric Constant and Permeability of a Dilute Suspension", *Philips Res. Rept.*, Vol. 30, p. 83-90, 1975

[48] R. Kress,
Linear Integral Equations, Springer, 1989

[49] E. M. Landesman, A.C. Lazer
"Nonlinear Perturbations of Linear Elliptic Boundary Value Problems at Resonance", *J. Math. Mech.*, Vol. 19, p. 609-623, 1970

[50] E. M. Landesman, A.C. Lazer
"Linear Eigenvalues and a Nonlinear Boundary Value Problem", *Pacif. J. Math.*, 33 (1970), 311–328.

[51] A.C. Lazer,
"A Second Look at the First Result of Landesman-Lazer Type", *J. Diff. Eqns.*, Vol.5, p. 113-119, 2000

[52] P. Li, S. Yau,
"On the Schrödinger Equation and the Eigenvalue Problem", *Commun. Math. Phys.*, Vol. 88, p. 309-318, 1983

[53] F. Loos,
"Modelling, Simulation and Optimization of Joule Heating Problems in Context with Electrical Devices and Connections", *Thesis*, to appear 2013

[54] M. Luskin,
"A Galerkin Method for Nonlinear Parabolic Equations with Nonlinear Boundary Conditions", *Siam J. Numer. Anal.*, Vol. 16, No. 2, 1979

[55] V. Marcenko, E. Khruslov,
Homogenization of Partial Differential Equations, Birkhäuser, 2007

[56] S. Martinez, J. D. Rossi, "Isolation and Simplicity for the First Eigenvalue of the p-Laplacian with a Nonlinear Boundary Condition", *Abstract and Applied Analysis*, Vol. 7, p. 287-293, 2002

[57] J. Mawhin and J. R. Ward,
"Nonresonance and Existence for Nonlinear Elliptic Boundary Value Problems", *Nonlinear Analysis, Theory, Methods and Applications*, 6 (1981), 677–684.

[58] J. Mawhin, J. R. Ward Jr., M. Willem
"Variational Methods and Semi-linear Elliptic Equations", *Arch. Rational Mech. Anal.*, Vol. 95, No. 3, p. 269-277, 1983

[59] V. Maz'ya, T. O. Shaposhnikova,
Theory of Sobolev Multipliers, Springer, 2009

[60] W. McLean,
Strongly Elliptic Systems and Boundary Integral Equations, Cambridge University Press, 2000

[61] I. Miyadera,
Nonlinear Semigroups, AMS Mathematical Monographs, Vol. 109, 1991

[62] J. R. Munkres,
Topology, Prentice Hall, 2000

[63] F. Murat, J. Simon
"Etude de Problemes d'Optimal Design" *Lecture Notes in Computer Science*, Vol. 41, p. 54-62, 1976

[64] J. Naumann,
Einführung in die Theorie parabolischer Variationsungleichungen, Teubner Verlagsgesellschaft, 1984

[65] G. Of, T. Phan Xuan, O. Steinbach,
"Boundary Element Methods for Dirichlet Boundary Control Problems", *Berichte aus dem Institut für Numerische Mathematik*, TU Graz, 2010/11

[66] L.E. Payne, H.F. Weinberger,
"An Optimal Poincaré Inequality for Convex Domains", *Arch. Rational Mech. Anal.*, Vol. 5, p. 286-292, 1960

[67] G. Pólya,
"On the Eigenvalues of Vibrating Membranes", *Proc. London Math. Soc.*, p. 419-433, 1961

[68] R. Potthast,
"Fréchet differentiability of boundary integral operators in inverse acoustic scattering", *Inverse Problems*, Vol. 10, p. 431-447, 1994

[69] P. Quittner, P. Souplet
Superlinear Parabolic Problems, Springer, 2007

[70] G. D. Raithby, K. G. T. Hollands,
"A General Method of Obtaining Approximate Solutions to Laminar and Turbulent Free Convection Problems", *Adv. Heat Transfer*, Vol. 11, p. 265-315, 1975

[71] W. Rudin,
Real and Complex Analysis, McGraw-Hill, 1987

[72] K. Ruotsalainen, W. Wendland,
"On the Boundary Element Method for some Nonlinear Boundary Value Problems", *Numer. Math.*, Vol. 53, p. 299-314, 1988

[73] C. Schwab,
"Variable Order Composite Quadrature of Singular and Nearly Singular Integrals", *Computing*, Vol. 53, p. 173-194, 1994

[74] D. Shi-qiang, L. Jia-chun,
"Homogenized Equations for Steady Heat Conduction in Composite Materials with Dilutely-Distributed Impurities", *Appl. Math. Mech.*, Vol. 4, No. 2, 1983

[75] R. Siegel, J. R. Howell,
Thermal Radiation Heat Transfer, Taylor & Francis, New York, 2002

[76] G. Teschl,
Ordinary Differential Equations and Dynamical Systems, American Mathematical Society, Providence, 2012

[77] H. Weyl,
"Über die Asymptotische Verteilung der Eigenwerte" *Nachrichten von der Gesellschaft der Wissenschaften zu Göttingen, Mathematisch-Physikalische Klasse*, 1911

[78] X. Xiong, C. Fu, H. Li,
"Central Difference Method of a Non-standard Inverse Heat Conduction Problem for Determining Surface Heat Flux from Interior Observations", *J. Applied Mathematics and Computation*, Vol. 173, p. 1265-1287, 2006

[79] E. Zeidler,
Nonlinear Functional Analysis and its Applications, I, Springer, 1985

[80] E. Zeidler,
Nonlinear Functional Analysis and its Applications, II/B, Springer, 1990

i want morebooks!

Buy your books fast and straightforward online - at one of world's fastest growing online book stores! Environmentally sound due to Print-on-Demand technologies.

Buy your books online at
www.get-morebooks.com

Kaufen Sie Ihre Bücher schnell und unkompliziert online – auf einer der am schnellsten wachsenden Buchhandelsplattformen weltweit! Dank Print-On-Demand umwelt- und ressourcenschonend produziert.

Bücher schneller online kaufen
www.morebooks.de

 VDM Verlagsservicegesellschaft mbH
Heinrich-Böcking-Str. 6-8
D - 66121 Saarbrücken
Telefon: +49 681 3720 174
Telefax: +49 681 3720 1749
info@vdm-vsg.de
www.vdm-vsg.de

Printed by Books on Demand GmbH, Norderstedt / Germany